Analysis Report on
Fire Fighter Fatalities

Prepared by

Fire Analysis and Research Division
National Fire Protection Association
Batterymarch Park
Quincy, MA 02269

Prepared for

Federal Emergency Management Agency
U.S. Fire Administration
Contract No. EMW-87-C-2570

August 1988

ACKNOWLEDGEMENTS

This study was funded by the U.S. Fire Administration of the Federal Emergency Management Agency (FEMA). It would not have been possible without the cooperation and assistance of the U.S. fire service, the United States Fire Administration, the Public Safety Officers' Benefits Program of the Department of Justice, the Bureau of Land Management of the Department of the Interior, the Bureau of Indian Affairs, the U.S. Department of Energy and the U.S. military, all of whom contributed data to this study.

The assistance of other NFPA staff members is also acknowledged: A. Elwood Willey, Assistant Vice President, Research and Fire Information Services; Dr. John Hall, Director, Fire Analysis and Research Division; Martin Henry, Director, Public Fire Protection Division; Kenneth Tremblay, John Barry, Michael Karter, and Kenneth Taylor, Fire Analysis and Research Division; and Gary Tokle and Carl Peterson, Public Fire Protection Division.

The authors also wish to acknowledge the assistance of USFA staff, particularly John Ottoson, Project Officer, for their suggestions and support.

Special thanks go to Nancy Schwartz and Helen Columbo for typing the several drafts of this report.

Rita F. Fahy
Arthur E. Washburn
Paul R. LeBlanc
Project Staff

TABLE OF CONTENTS

Page

Table of Contents i

List of Figures ii

List of Tables iv

Background V

I.   Introduction I

A.   Who Is a Fire Fighter?. 1
B.   What Constitutes an On-Duty Fatality? 2
C.   Sources of Initial Notification 3
D.   Procedure for Including a Fatality in the Study . 4
E.   Additional Data Collection Completed
        for the Contract. 5

II.   1987 Findings 6

A.   Type of Duty. 6
B.   Cause and Nature of Fatal Injury or Illness. . 10
C.   Ages of Fire Fighters 14
D.   Fire Ground Deaths. 17
E.   Time of Day 19
F.   Month of the Year 19
G.   State and Region. 24
H.   Analysis of Urban/Rural/Suburban Patterns
        in Fire Fighter Fatalities. 24

III. Training Fatalities 1978-1987 29

IV.   Wildland Fire Fatalities 1978-1987. 35

V.   Smoke Inhalation and Smoke Exposure Deaths 1978-1987. 43

VI.   Conclusions and Recommendations 49

      References 52

LIST OF FIGURES

Figure	Title	Page
1	Line of Duty Fire Fighter Deaths 1978-1987	7
2	Average Annual Fatality Rates by Industry, 1980-1984	8
3	Fire Fighter Deaths 1987 by Type of 'Duty	9
4	Fire Fighter Deaths 1987 by Cause of Injury	12
5	Fire Fighter Deaths 1987 by Nature of Injury	13
6	Fire Fighter Deaths 1987 by Age and Cause of Death	15
7	Average Death Rates per 10,000 Fire Fighters 1983-1987	16
8	Fire Ground Deaths in 1987 by Fixed Property Use	18
9	Fire Fighter Fatalities 1987 by Time of Day	20
10	Fire Fighter Fatalities by Time of Day 1978-1987	21
11	Fire Fighter Fatalities 1987 by Month of Year	22
12	Fire Fighter Fatalities by Month of Year 1978-1987	23
13	U.S. Fire Fighter Training Deaths 1978-1987	30
14	U.S. Fire Fighter Training Deaths by Cause of Injury 1978-1987	32
15	U.S. Fire Fighter Training Deaths by Nature of Injury 1978-1987	33
16	U.S. Fire Fighter Deaths in Wildland Fires, 1978-1987	36
17	Wildland Fire Ground Fatalities by Region 1978-1987	37

LIST OF FIGURES (Continued)

<u>Figure</u>	<u>Title</u>	<u>Page</u>
18	Fire Fighter Fatalities in Wildland Fires by Cause of Fatal Injury 1978-1987	38
19	Smoke Inhalation and Smoke Exposure Fire Ground Deaths 1978-1987	44
20	Causes of Smoke Inhalation Deaths 1978-1987	48

LIST OF TABLES

Table	Title	Page
1	Fire Fighter Fatalities by State	25
2	Fire Fighter Death Rates by Region 1987	26
3	U.S. Wildland Fire Fighter Fatalities by Nature of Fatal Injury 1978-1987	41
4	Causes of Smoke Inhalation Deaths 1978-1987	47

BACKGROUND

For more than a decade, the National Fire Protection Association (NFPA) has developed the most complete records on U.S. fire fighter fatalities - both in breadth of coverage and depth of detail - of any organization. This data base has been used to support the fatality studies produced each year by NFPA since 1974.

Over the past seven years, NFPA also has worked with FEMA's U.S. Fire Administration (USFA) to provide, in a timely manner, lists of fire fighter fatalities and their next of kin to support the National Fire Academy's annual Fire Fighter Memorial Service. Under the present contract, NFPA has provided the USFA with lists, both hand lettered and typed, of 1987 fire fighter fatalities and with a list of names and addresses of next of kin and of fire department chiefs for use in the Memorial Service in October 1988.

In August, a briefing on the 1987 experience and three special analyses was presented by NFPA staff to USFA staff in Emmitsburg, MD. Through the briefing and analysis, this contact continued the trend toward more extensive analysis of patterns and trends in specific parts of the fire fighter fatality problem. With over a decade of experience now classified in a computer data base, NFPA is able to provide increasingly detailed and focused examinations of the specific parts of the problem addressable by particular strategies.

The deliverables under this contract are (a) this analysis report, (b) the incident and casualty data on tape in NFIRS Version 4.0 format, which is being delivered separately, (c) the various lists described above, and (d) the briefing provided in August.

I. INTRODUCTION

The purpose of this study is to analyze the circumstances surrounding fire fighter fatalities in the United States in 1987 in an attempt to identify potential means for reducing the number of deaths that occur each year. In addition to the 1987 findings, this study will also include special analyses of particular recurring scenarios, using NFPA's data base of fire fighter fatalities from 1978 through 1987.

A. Who Is a Fire Fighter?

For the purpose of this study, the term "fire fighter" covers all members of organized fire departments, whether career, volunteer or mixed; full-time public service officers acting as fire fighters; state and federal government fire service personnel; temporary fire suppression personnel operating under official auspices of one of the above; and privately employed fire fighters including trained members of industrial or institutional fire brigades, whether full- or part-time.

Under this definition, the study includes not just municipal fire fighters, but also seasonal and full-time employees of the U.S. Forest Service and state forestry agencies; prison inmates serving on state forest service crews; fire fighters for the Bureau of Land Management, the Bureau of Indian Affairs, and the U.S. Department of Energy; military personnel performing assigned fire suppression activities; civilian fire fighters working at military installations; and members of industrial fire brigades.

B. What Constitutes an On-Duty Fatality?

The term "on-duty" refers to being at the scene of an alarm, whether a fire or non-fire incident; being enroute while responding to or returning from an alarm; performing other assigned duties such as training, maintenance, public education, inspection, investigations, court testimony and fund raising; performing non-fire duties on official assignment: and being on call, under orders or on stand-by duty other than at home or at the individual's place of business.

On-duty fatalities include any injury sustained while on duty that proves fatal, any illness that was incurred as a result of actions while on duty that proves fatal, and fatal mishaps involving occupational hazards that occur while on duty. The types of injuries included in the first category are mainly those that occur on the fire ground, in training or in accidents while responding to or returning from alarms. The most common examples of fatal illnesses incurred on duty are fatal heart attacks. Another example is a fire fighter who contracted hepatitis when a victim being transported by ambulance pulled out his intravenous needle and stuck the fire fighter. A few examples of fatal occupational mishaps include fire fighters who died of asphyxiation while working on fire apparatus in closed garages, a fire fighter who fell through a slide pole hole, a fire fighter electrocuted while raising a banner for a town event, a volunteer fire fighter who was fatally injured when he fell down a flight of stairs in his home while responding to an alarm, and a fire inspector who fell through a skylight.

Also included in the study are fire fighters who were murdered while on duty. These include fire fighters shot by snipers while on the fire ground and fire fighters shot in the station by off-duty or former fire fighters.

Fatal injuries and illnesses are included even In cases where death is considerably delayed. When the onset of the condition and death occur in different years, the incident is counted on the basis of the former. For example, a Michigan fire fighter died in 1986 of a brain injury received in 1979 when he was struck by a brass hose coupling, resulting in recurring seizures. Because his death was the direct result of his injury, and the injury occurred in 1979, he is counted as a 1979 fatality.

The NFPA recognizes that these definitions should include chronic illnesses (such as cancer) that prove fatal and that arise from occupational factors. In practice, there is as yet no mechanism for identifying fatalities that are due to illnesses that develop over long periods of time and that thereby present an ambiguous picture on the issue of occupational versus other factors as causes. This is recognized as a gap that cannot now be filled because of the limitations of the state of the art in tracking and analysis.

C. Sources of Initial Notification

As an integral part of its ongoing program to collect and analyze fire data, NFPA solicits information on fire fighter fatalities from the U.S. fire service and a wide range of other sources. These include the U.S. Fire Administration and the Public Safety Officers' Benefits Program (PSOB). Both are organizations with whom NFPA has maintained long-standing cooperative efforts in collecting and analyzing fire fighter fatality data. Other contacts include federal agencies such as the U.S. Forest Service of the Department of Agriculture, the Bureau of Indian Affairs and the Bureau of Land Management of the Department of Interior, the U.S. military, the Department of Energy, and the Occupational Safety and Health Administration (OSHA).

We also receive notification from fire service organizations such as the International Association of Fire Fighters, state fire associations, state training organizations, state and local fire marshals, and fire service publications. A network developed over the years of individuals interested in the area of fire fighter fatalities also assists in identifying incidents, especially those that occur outside of large urban areas or that involve non-fire-incident-related fatalities. Among these individuals are fire fighters, photographers, fire buffs, and members of the insurance industry.

Notification of fatal incidents also comes from NFPA members and staff and through the use of a newspaper clipping service that reads all daily and weekly newspapers in the country.

D. Procedure for Including a Fatality in the Study

After initial notification of a fatal incident is received, contact with the local fire department is made by telephone to verify the incident, its location and the fire department involved. Data collection forms for the fatality and the fire, if it was a fire incident, are sent to the responsible local official identified during the telephone follow-up. After the forms are returned to NFPA, a final decision is made to include or exclude the fatality, based on the inclusion criteria described previously. In order to make a final determination, additional information is sometimes sought, either by contacting the fire department directly to clarify some of the details or by obtaining data elsewhere, such as medical documentation frequently available from PSOB.

Some of the material that might be received to document an incident includes casualty forms, both NFPA fire fighter fatality study reporting forms and NFIRS-type forms; NFPA's Fire Incident Data Organization major-fire report form or the department's own incident reporting form, if a fire incident was involved in the fatality; medical data such as death certificates or autopsy reports: special 'investigation reports from other agencies; police and motor vehicle accident reports, if applicable; photographs and diagrams; and additional newspaper accounts. Incidents to be included in the study are then coded into NFPA's Fire Incident Data Organization (FIDO), which includes both incident and casualty information. By mutual agreement of the USFA and NFPA project staff, the same inclusion criteria were used for the USFA study as are used in the NFPA study.

Work described to this point was done as part of NFPA's ongoing program of data collection and analysis in the area of fire fighter fatalities and was completed at no cost to FEMA.

E. Additional Data Collection Completed for the Contract

To meet FEMA's request for a list of the next-of-kin of the 1987 fatalities and the names and addresses of the fire chiefs, a follow-up mailing was sent to all departments asking them to verify the victims' names and dates of fatal injury, the names and addresses of the departments and chiefs, and the names and relationships of the next of kin. Telephone calls were made to non-responding fire departments to obtain the information.

One hundred twenty-seven fire fighters died in the line of duty in 1987. As shown in Figure 1, although this was an 11.4 percent increase from the year before, there has been no consistent trend overall since 1981. Renewed efforts must be made to target certain specific areas of the problem in order to achieve further reduction. A comparison of fire fighting to other occupations is shown in Figure 2. The data for other industries was obtained from a NIOSH report.' The fire fighter fatality data was based on deaths of career fire fighters reported to NFPA from 1980 to 1984 and NFPA's estimate of the number of career fire fighters in the United States in 1983. After mining, fire fighting is the most hazardous occupation in terms of fatalities. This study will report some of the most frequently occurring scenarios and will present some conclusions and recommendations to address the problem.

A. Type of Duty

The distribution of deaths by type of duty being performed is shown in Figure 3. The largest proportion of deaths occurred during fire ground operations. Of these 53 deaths, 24 were due to heart attacks, 10 to internal trauma, 8 to smoke inhalation, six to crushing injuries, two to burns, two to gunshot wounds and one to electrocution. Twenty-four of the victims were career fire fighters and 29 were volunteers.

The second largest category involved responding to and returning from alarms, which accounted for over a quarter of the deaths. This is an improvement over the experience in 1986 when one-third of the deaths occurred while responding or returning. Sixteen of these 37 deaths were due to heart

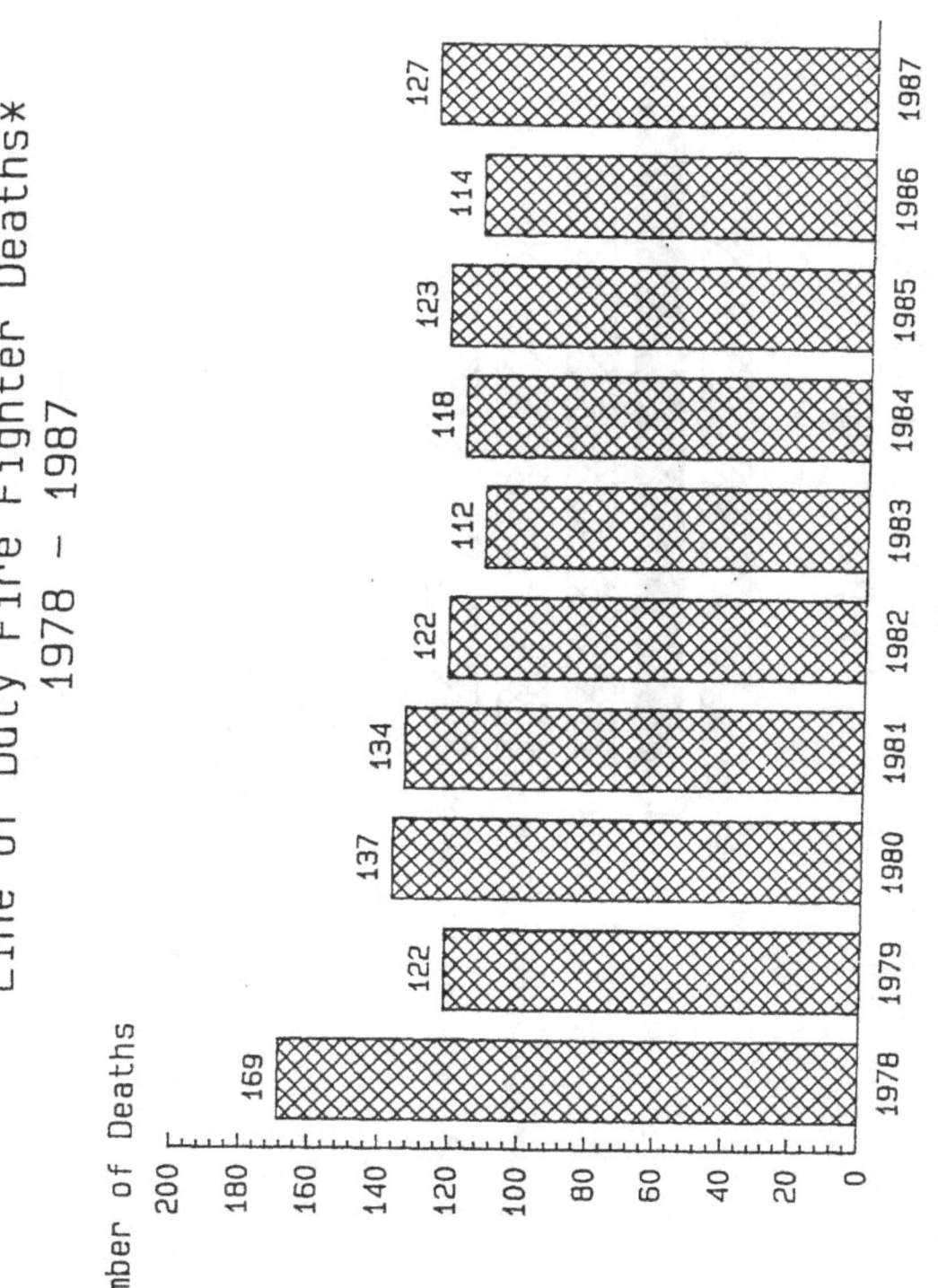

Figure 1
Line of Duty Fire Fighter Deaths*
1978 – 1987

* The number of fire fighter deaths may differ
from previously reported numbers because of
information received since publication.

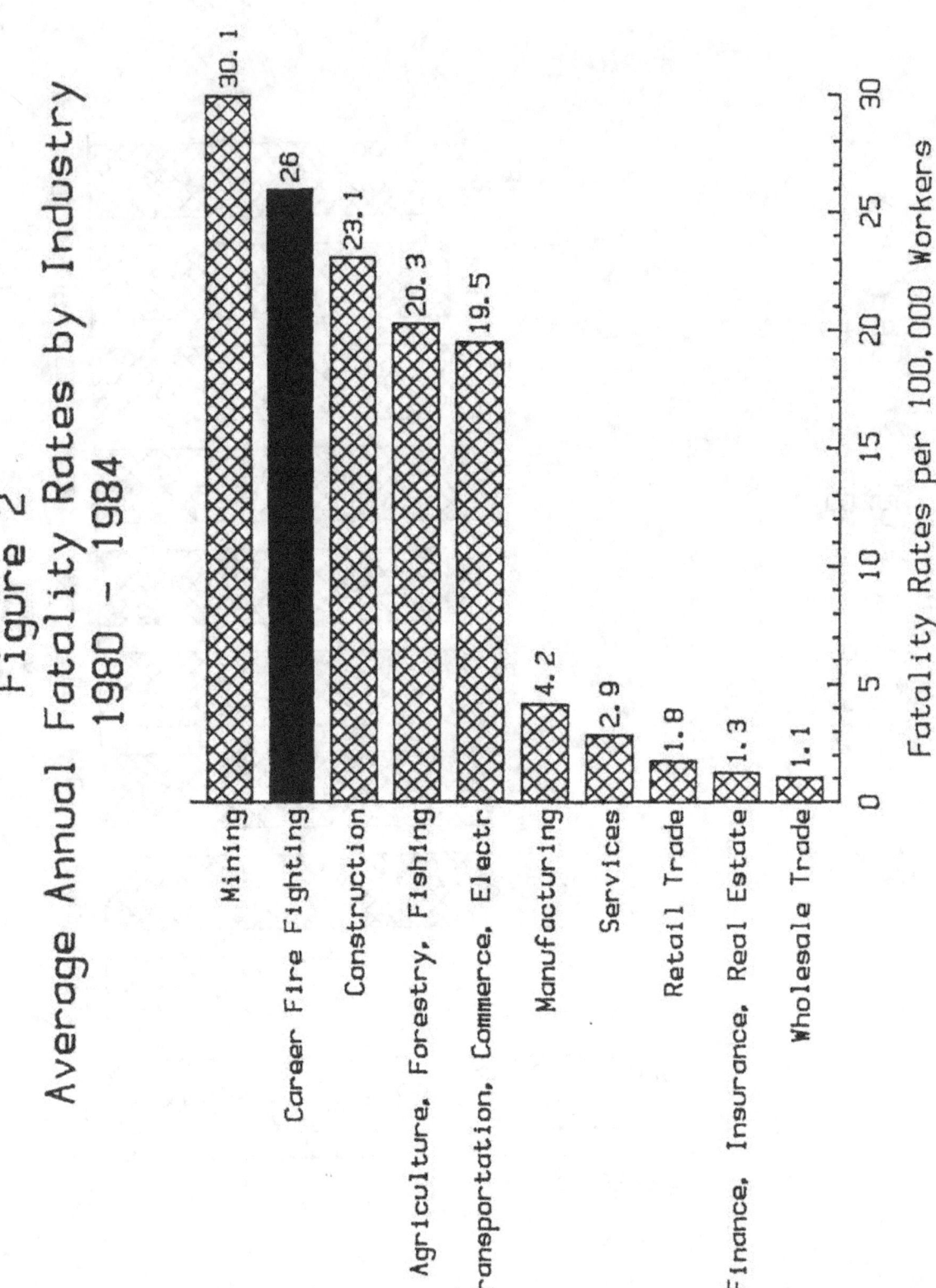

Figure 2

Average Annual Fatality Rates by Industry
1980 – 1984

Mining: 30.1
Career Fire Fighting: 26
Construction: 23.1
Agriculture, Forestry, Fishing: 20.3
Transportation, Commerce, Electr: 19.5
Manufacturing: 4.2
Services: 2.9
Retail Trade: 1.8
Finance, Insurance, Real Estate: 1.3
Wholesale Trade: 1.1

Fatality Rates per 100,000 Workers

0 5 10 15 20 25 30

Source: Fire fighter fatality data from
NFPA. All other data from NIOSH.

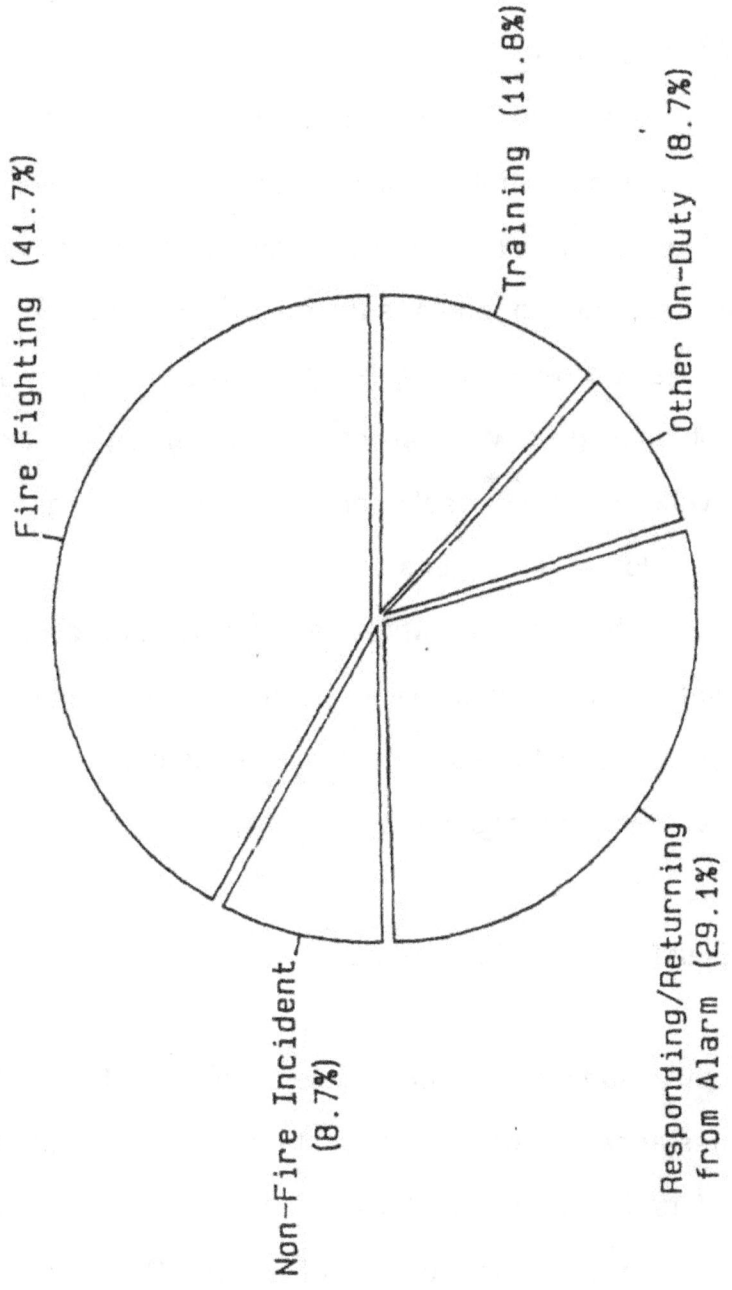

Figure 3
Fire Fighter Deaths 1987
by Type of Duty

Fire Fighting (41.7%)

Training (11.8%)

Other On-Duty (8.7%)

Responding/Returning from Alarm (29.1%)

Non-Fire Incident (8.7%)

attacks, five to falls from apparatus, one to burns, one to heat stroke, two to trees falling on vehicles and the rest to collisions.

There were 15 training-related fatalities last year, compared to seven in 1986. Four of the deaths involved participants in two live fire training exercises. Because of the unusual increase in the number of training deaths in 1987, this report includes a separate section that discusses deaths during training from 1978 through 1987.

There were 11 deaths while working at non-fire incidents. These included four heart attacks and one stroke while working at motor vehicle accidents or EMS calls, two fire fighters struck by passing vehicles while working at motor vehicle accidents, one due to burns suffered in an explosion while working at a natural gas line leak, one drowning after a water rescue, one victim who fell off the side of an engine while working at an ammonia leak and was crushed by the vehicle's rear wheels, and one while diving to retrieve nets and rigging from a lost fishing vessel.

The other 11 deaths occurred while performing other duties -- 7 deaths during normal station and administrative duties, one during a fire inspection, one during maintenance activities, one while assisting at a water reservoir cleanup, and one while participating in a parade.

B. Cause and Nature of Fatal Injury or Illness

As used in this study, the term "cause" refers to the action, lack of action, or circumstances that directly resulted in the fatal injury while the term "nature" refers to the medical nature of the fatal injury or illness or what is often referred to as the cause of death. Often, the fatal injury is the result of a chain of events, the first of which is recorded as the cause. For example, if a fire fighter is struck by a collapsing wall, becomes trapped by the debris, runs out of air before being rescued and dies of asphyxiation, the cause of fatal injury recorded is "struck by collapsing wall" and the nature of fatal injury is "asphyxiation."

Figure 4 shows the distribution of deaths by cause of fatal injury or illness. Stress was reported as the cause in almost half of the deaths. Eight of these 62 deaths were attributed to strenuous physical activities. The second major category was struck by or contact with objects. These 32 deaths included 24 motor vehicle accidents, three fire fighters struck by collapsing walls, three struck by falling objects and two who were fatally shot on arrival at the scene of a structure fire.

Nineteen fire fighters were caught or trapped - six by fire progress, five by interior collapse, three by being lost inside buildings, two by underwater objects and one each by flashover, explosion and falling objects. Ten fire fighters died as a result of falls - seven from fire apparatus, one through a hole burned in the floor of a building, one from a structure, and one who tripped or lost his balance and fell, striking his head. Four deaths were caused by exposure to fire products - three to smoke and one to fire.

Figure 5 shows the distribution of deaths by the medical nature of the fatal injury or illness. The largest proportion of deaths were due to heart attacks. Of these 59 deaths, medical documentation indicated that 14 of the victims had prior heart problems, either previous heart attacks or bypass surgery, and 12 others had severe arteriosclerotic heart disease (defined for this study as arterial occlusion of at least 50 percent). Two other victims suffered from hypertension and another was fatigued prior to the fatal incident. Medical documentation was not available for the other 30 heart attack victims.

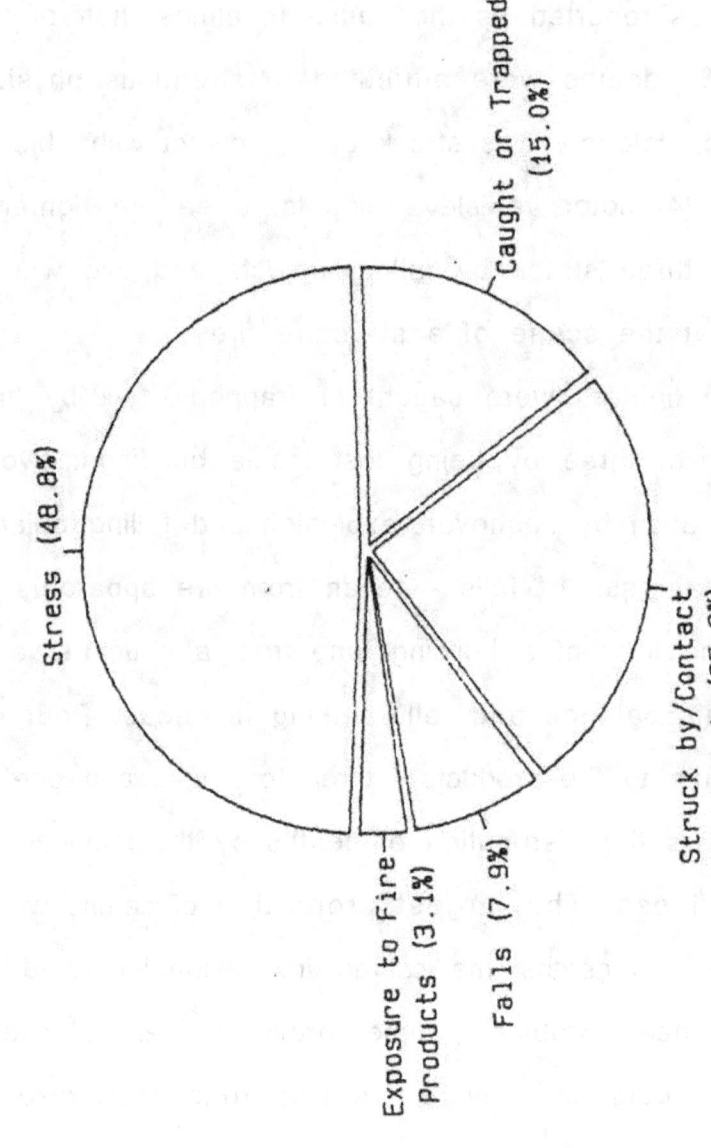

Figure 4
Fire Fighter Deaths 1987
by Cause of Injury

Caught or Trapped
(15.0%)

Stress (48.8%)

Struck by/Contact
with Object (25.2%)

Exposure to Fire
Products (3.1%)

Falls (7.9%)

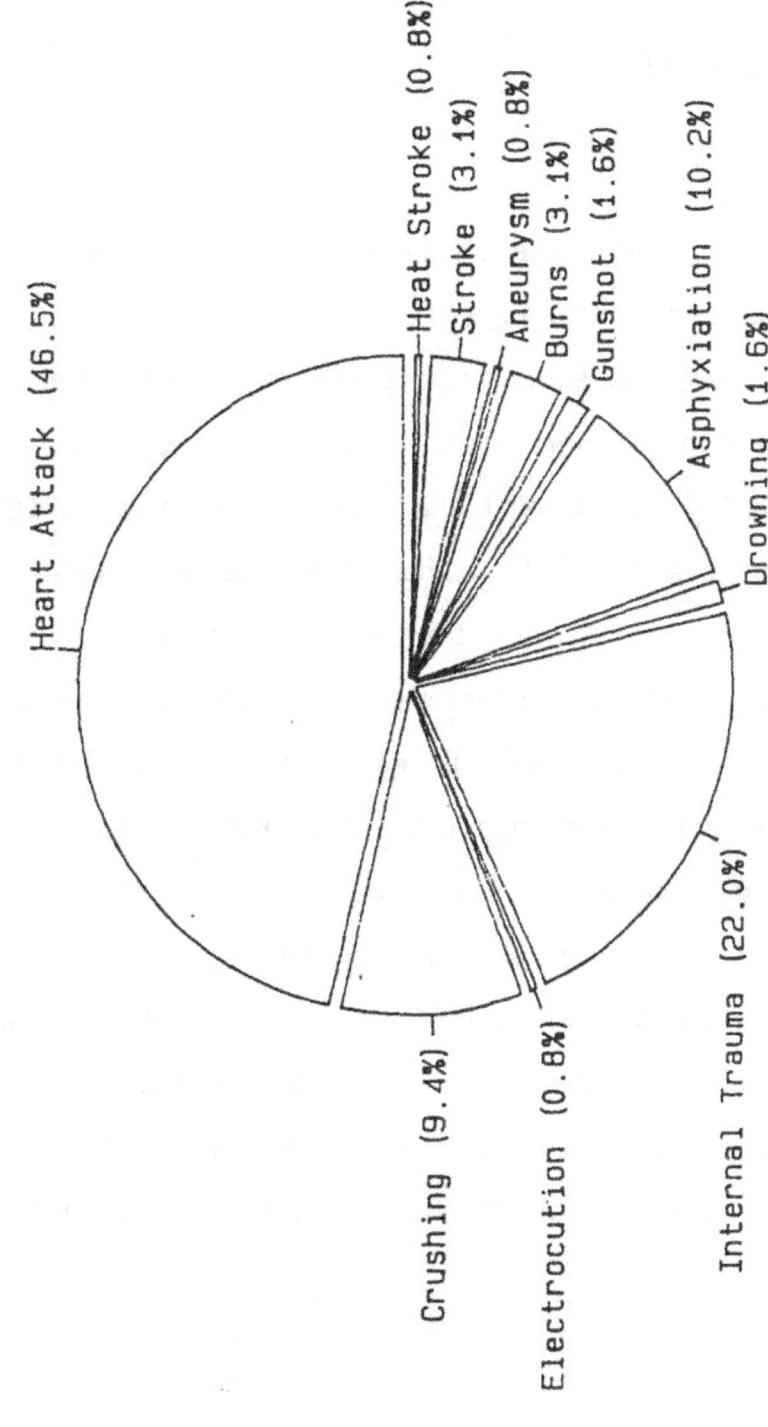

Figure 5
Fire Fighter Deaths 1987
by Nature of Injury

Heart Attack (46.5%)

Heat Stroke (0.8%)
Stroke (3.1%)
Aneurysm (0.8%)
Burns (3.1%)
Gunshot (1.6%)
Asphyxiation (10.2%)
Drowning (1.6%)

Crushing (9.4%)
Electrocution (0.8%)
Internal Trauma (22.0%)

The other categories of nature of fatal injury were internal trauma (28 deaths), asphyxiation (13 deaths), crushing (12 deaths), stroke (four deaths), burns (four deaths), drowning (two deaths), gunshot wounds (two deaths), and one each due to aneurysm, electrocution and heat stroke.

C. Ages of Fire Fighters

The ages of fire fighters who died in 1987 ranged from 16 to 73 years with a median age of 45 years. The distribution of fire fighter deaths by age and cause of death is displayed in Figure 6.

In most years, we find a very high proportion of heart attack deaths among older fire fighters and few heart attacks among fire fighters under age 35. In 1987, however, the rate of heart attack deaths was down to 60.5 percent among fire fighters over age forty (compared to over 75 percent in 1986), and less than half of the fire fighters over age 60 died of heart attacks. Five of the heart attack victims were under the age of 35; two were under 20.

Figure 7 shows the death rates by age categories using estimates of the number of fire fighters in each age group from NFPA's 1985 profile of fire departments and the fatality data from 1983 through 19873. As the graph shows, the death rate is lowest for fire fighters aged 20 to 39, slightly above the average rate for those aged 40 to 49, and much higher than average for fire fighters aged 50 and over. This is a reflection of the fact that although only 15 percent of all fire fighters are over age 50, that age group accounted for over a third of the deaths from 1983 through 1987.

Figure 6
Fire Fighter Deaths 1987
by Age and Cause of Death

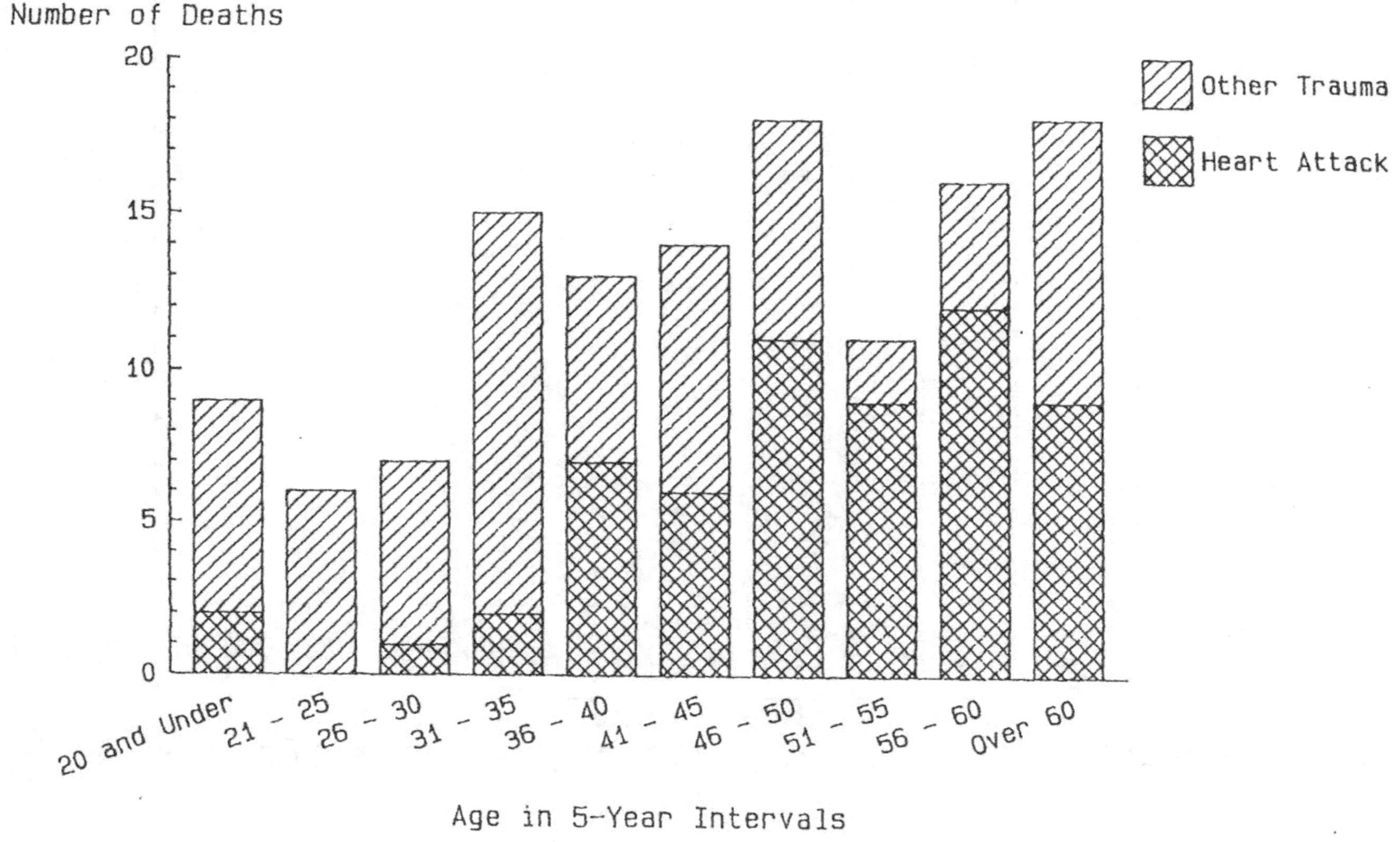

Number of Deaths
20
15
10
5
0
Other Trauma
Heart Attack
20 and Under
21 - 25
26 - 30
31 - 35
36 - 40
41 - 45
46 - 50
51 - 55
56 - 60
Over 60
Age in 5-Year Intervals

Figure 7

Average Death Rates per 10,000 Fire Fighters
1983 – 1987

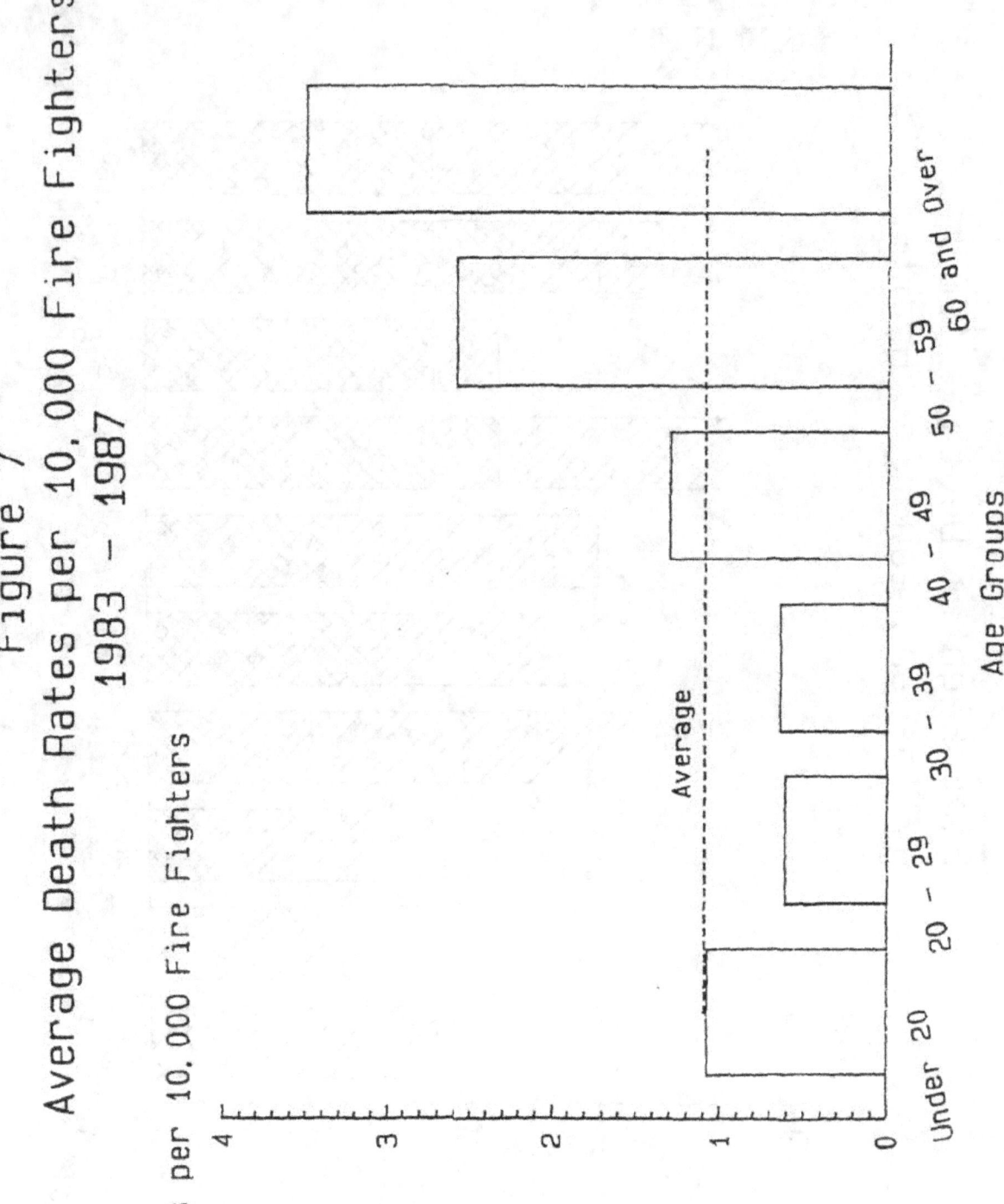

Deaths per 10,000 Fire Fighters

Age Groups

Note: These figures combine career and volunteer fire
fighters. The two groups may have very different age
distributions, which are not reflected here.

D. Fire Ground Deaths

This distribution of the 53 fire ground deaths by fixed property use is shown in Figure 8. For the first time in the 11 years of this study, the greatest proportion of fire ground deaths (32.1 percent) occurred in wildland fires - 17 deaths in 1987. (Another five deaths occurred while responding to or 'returning from such fires.) Wildland fire incidents in 1987 and over the past ten years are discussed in more detail in a separate section of the report.

The second largest proportion of deaths occurred in residential properties (24.5 percent) with nine in single-family dwellings, three in apartment buildings and one in a two-family dwelling. There were five deaths in storage properties - three in residential garages and two in barns. There were three deaths each in manufacturing properties and stores. There were two deaths in church fires and one in a college building.

Four fire fighters were killed in incendiary fires that originated in abandoned buildings. One of these fires, which involved a large warehouse complex, killed three Detroit, Michigan fire fighters in two separate situations. In that incident, one fire fighter fell from a third floor window after being forced to the window ledge by flashover while checking on the extent of fire involvement. The other two were killed several hours later in the same incident when a fire wall collapsed lapsed while they operated a hand line on the third story of an exposed and involved paper warehouse in the complex. (For a full description of the incident see "Triple Tragedy in Detroit" in the July 1987 issue of Fire Command.)

Two fatalities occurred in vacant buildings and one in a building under construction. These fires were also of incendiary origin. A suspicious fire in a building under renovation resulted in the death of one fire fighter.

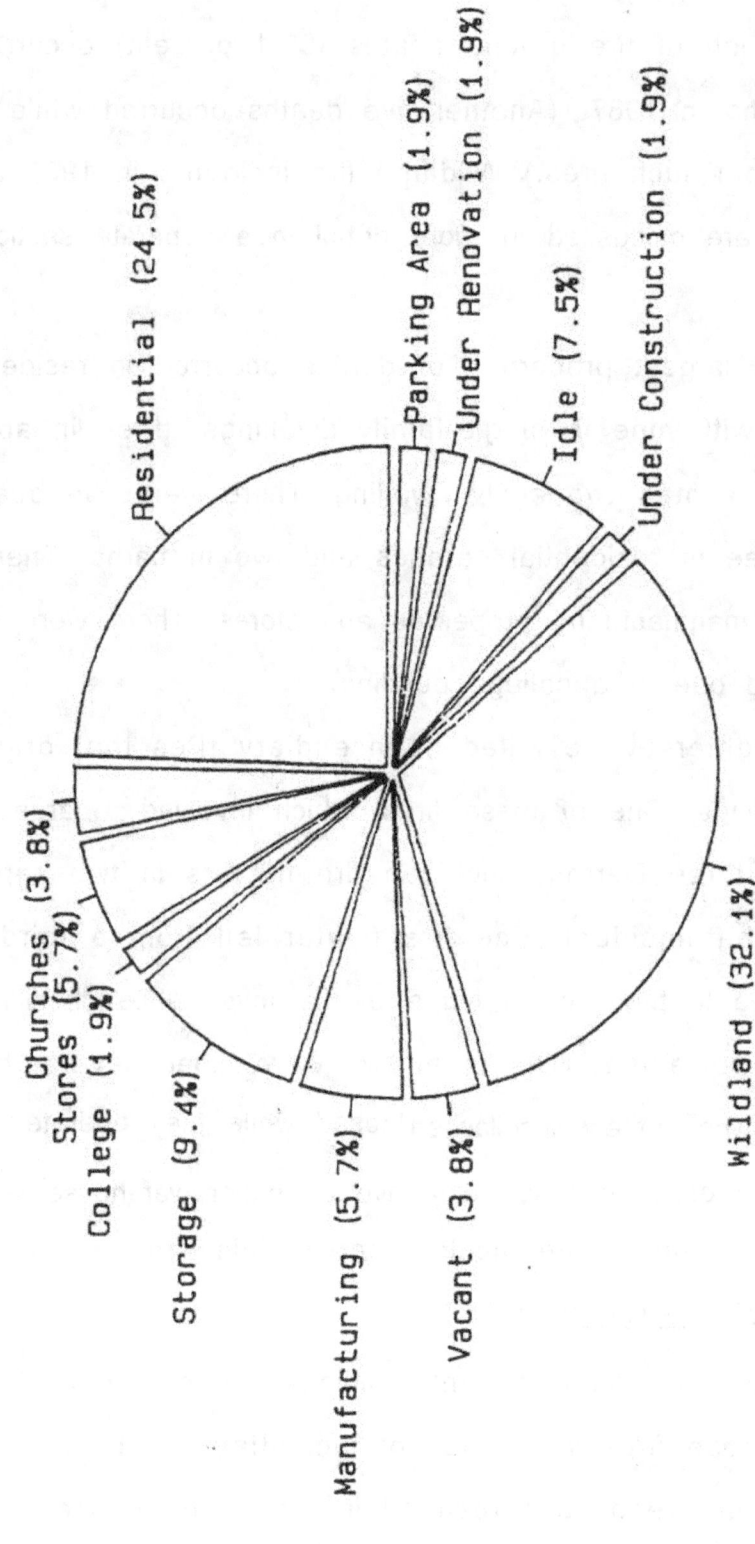

Figure 8
Fire Ground Deaths in 1987
by Fixed Property Use

Residential (24.5%)
Parking Area (1.9%)
Under Renovation (1.9%)
Under Construction (1.9%)
Idle (7.5%)
Wildland (32.1%)
Vacant (3.8%)
Manufacturing (5.7%)
Storage (9.4%)
College (1.9%)
Stores (5.7%)
Churches (3.8%)

A career chief suffered a heart attack while directing operations at a fire that erupted during the removal of a leaking underground fuel tank.

To put the hazards of fire fighting in various types of occupancies into perspective, the number of deaths per 100,000 structure fires was examined by fixed property use. The rates were calculated using the estimates of fire experience from NFPA's 1987 fire loss study4. There were 2.4 fire fighter deaths per 100,000 residential structure fires, compared to 11.7 deaths per 100,000 nonresidential structure fires. Although almost three times as many fires occurred in residential structures, the size, complexity and special hazards often associated with nonresidential structures result in a much greater risk at such fires.

E. Time of Day

The distribution of 1987 fire ground deaths and total deaths by time of alarm is shown in Figure 9. This graph shows a peak in the late afternoon for both categories. The number of deaths overall has two lower peaks in the late morning and late evening but otherwise is fairly constant throughout the day. The distribution of deaths by time of day over a ten-year period is shown in Figure 10. The number of deaths in both categories was at the highest between 1:00 and 9:00 pm and drops to the lowest level in the early morning hours.

F. Month of the Year

Figure 11 shows the distribution of 1987 fire fighter deaths by month. The same information for 1978 through 1987 is shown in Figure 12. Eleven of the deaths from late August to early October were associated with the series of forest fires that occurred in the western states. The ten-year analysis shows that fire ground deaths are higher in the winter months and in midsummer.;

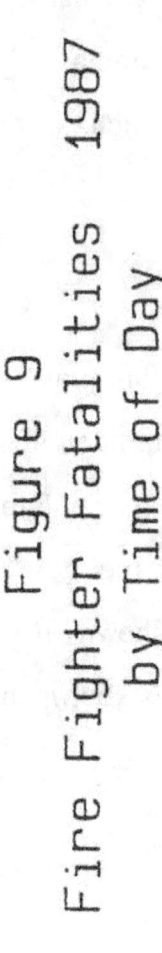

Figure 9
Fire Fighter Fatalities 1987
by Time of Day

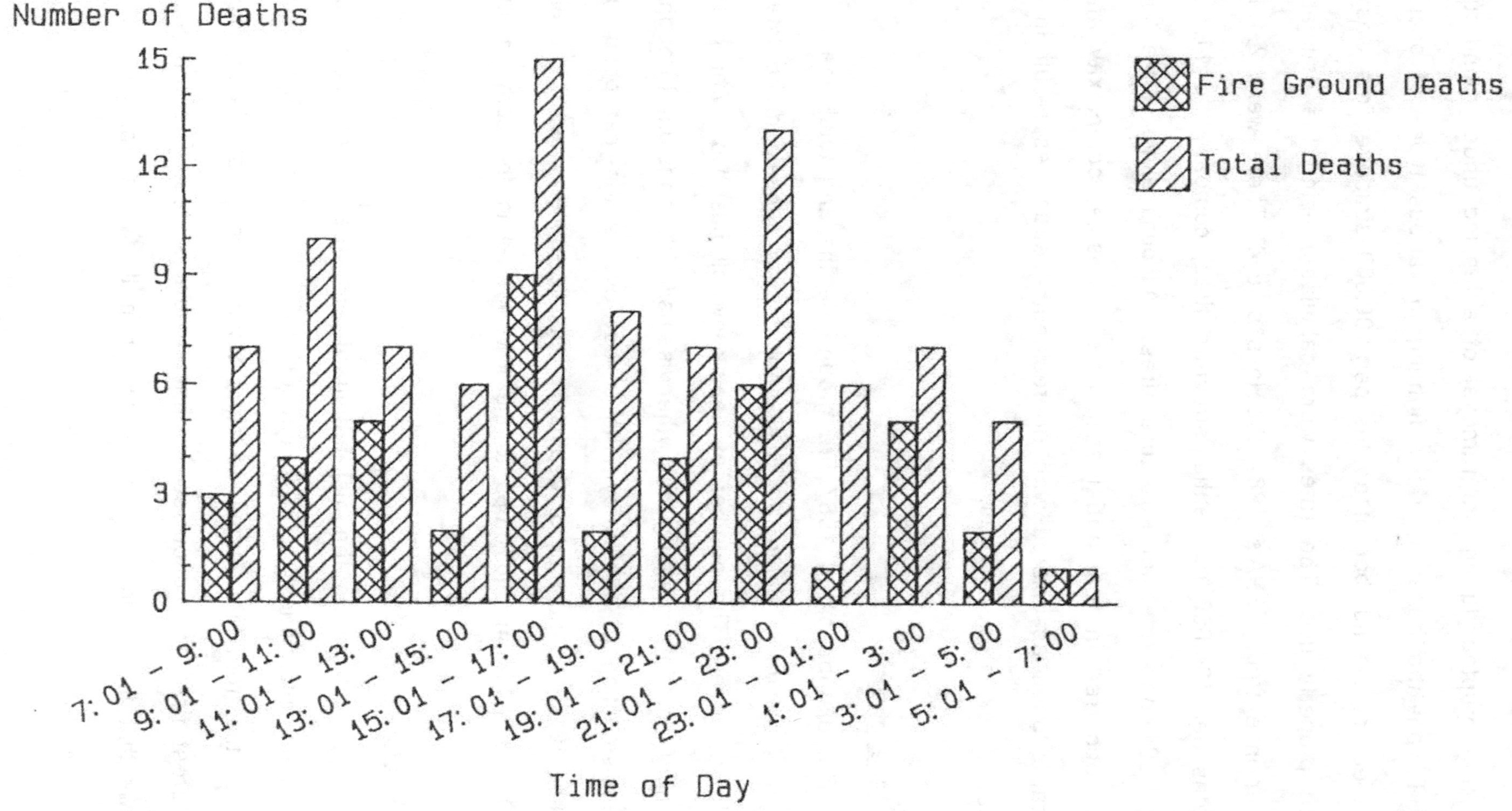

Based on 42 fire ground fatalities and
92 total fatalities for which time
was known.

Figure 10
Fire Fighter Fatalities
by Time of Day
1978 – 1987

Based on 598 fire ground fatalities and 1017 total fatalities for which time was known.

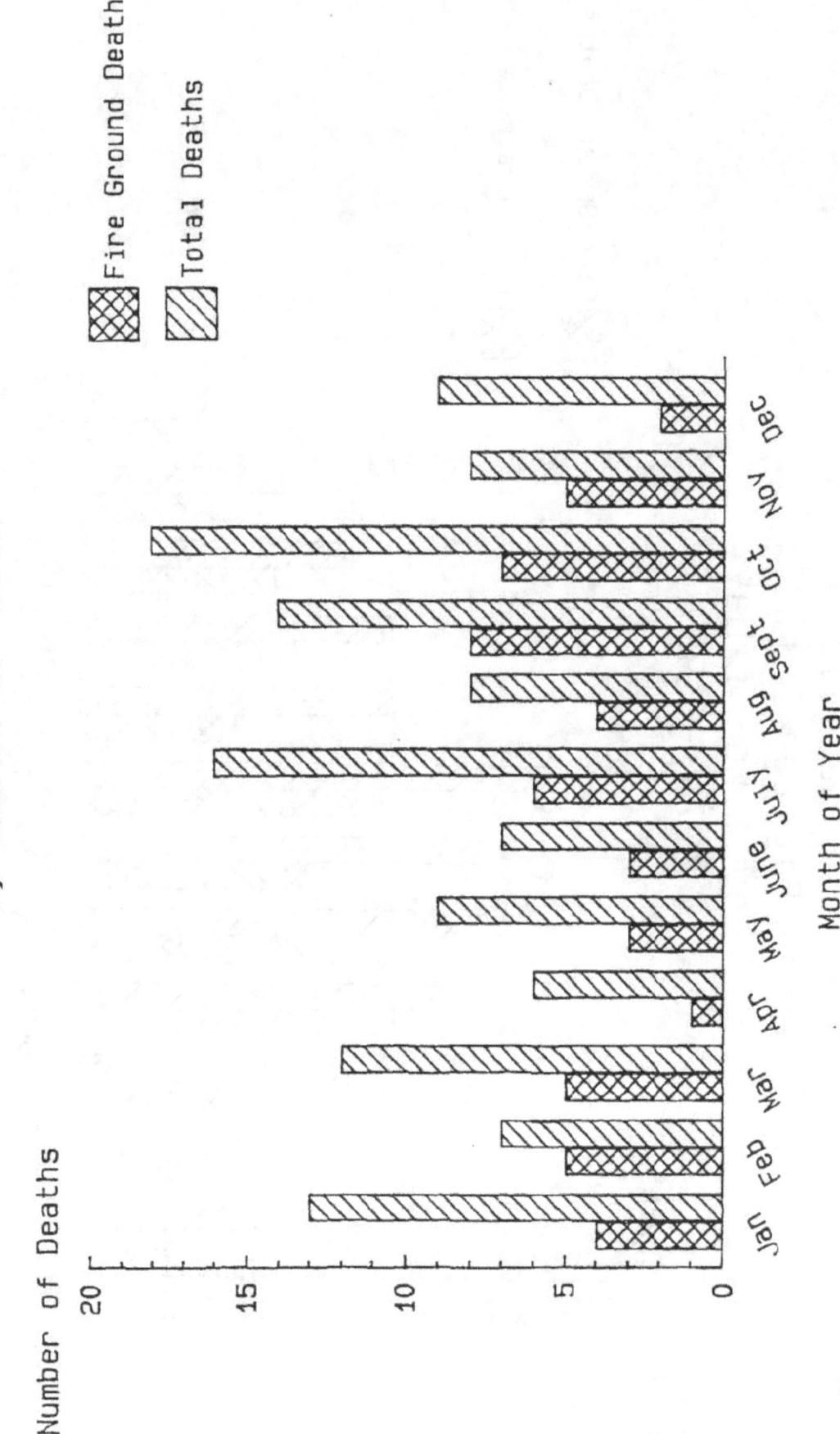

Figure 11
Fire Fighter Fatalities 1987
by Month of Year

Based on 53 fire ground fatalities and 127 total fatalities.

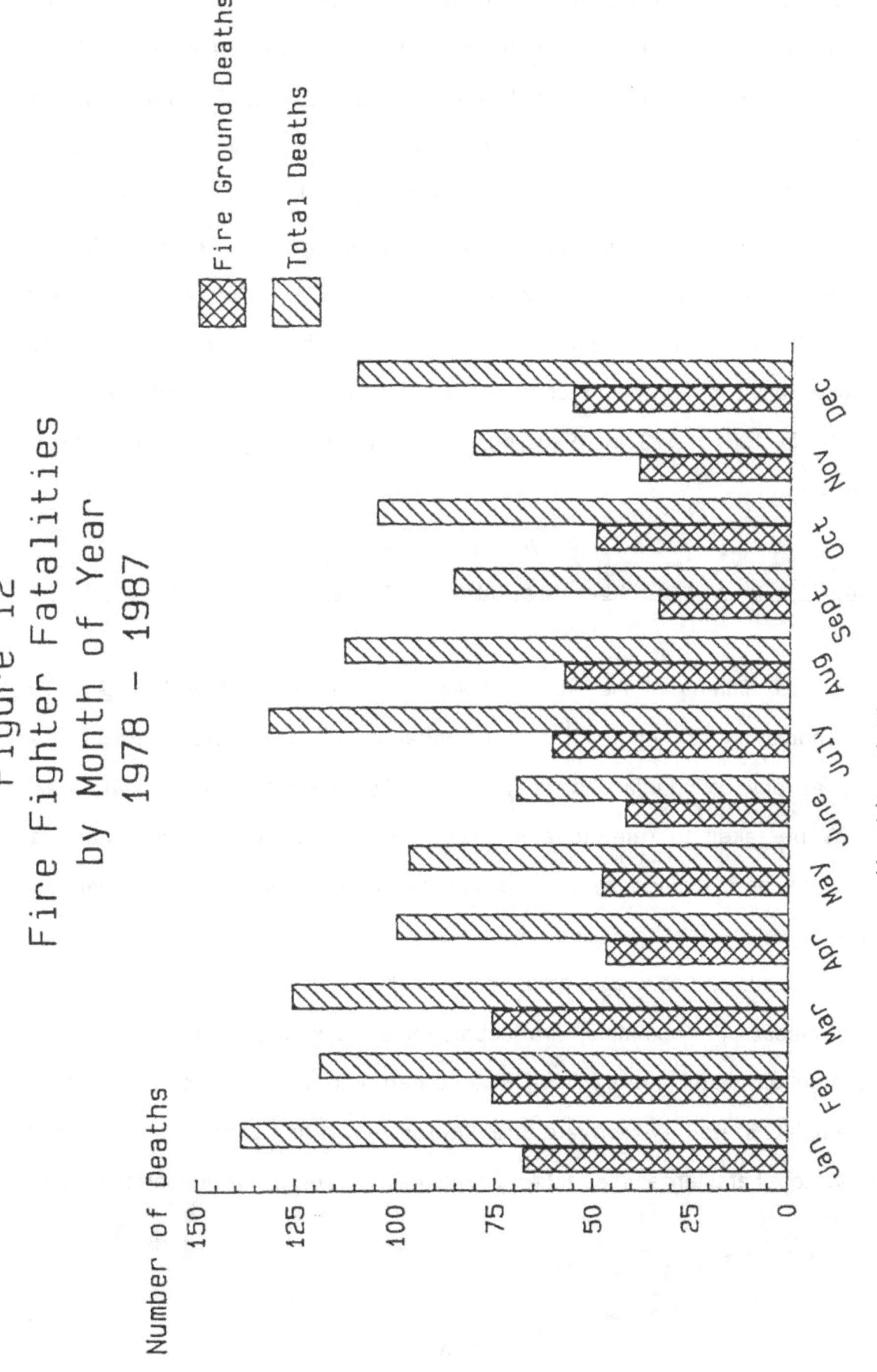

Figure 12
Fire Fighter Fatalities
by Month of Year
1978 – 1987

Based on 655 fire ground fatalities and
1278 total fatalities.

G. State and Region

The distribution of fire fighter deaths by state is shown in Table 1. Thirty-six states are represented on the list, led by New York with 15 deaths. The experience by region' is displayed in Table 2. The Northeast lost the largest number of fire fighters (44), followed by the South (36), the Northcentral region (25) and the West (22). When looking at fire ground deaths, we see that although there were more fire ground deaths in the Northeast, there were so many fewer fires in the West than in the Northeast that the death rate per 100,000 fires was highest in the West. 'This is not a surprising result given the much increased number of fire ground deaths in the West (14 in 1987 vs. 3 In 1986) due mostly to the wildland fire experience in California.

H. Analysis of Urban/Rural/Suburban Patterns in Fire Fighter Fatalities

The U.S. Bureau of the Census defines "urban" as a place having at least 2,500 population or lying within a designated urbanized area. "Rural" is defined as any community that is not urban. "Suburban" is not a Census term but may be taken to refer to any place, urban or rural, that lies within a metropolitan area defined by the Census but is not one of the designated central cities of that metropolitan area.

Fire department coverage areas do not always conform to the boundaries of Census places. For example, fire departments defined by counties or special fire protection districts may have both urban and rural sections, and there are Federal, state, and private fire fighters. In such cases, it may not be possible to characterize the entire coverage area of a fire department as

Table 1
1987 Line of Duty
Fire Fighter Fatalities

State	Number of Deaths		State	Number of Deaths
Alabama	2		Mississippi	2
Arizona	2		Missouri	2
California	13		Nebraska	1
Colorado	1		New Hampshire	4
Connecticut	4		New Jersey	7
District of Columbia	1		New Mexico	2
Florida	4		New York	15
Georgia			North Carolina	2
Illinois			Ohio	2
Indiana	3		Oregon	1
Kansas	5		Pennsylvania	7
Kentucky	2		South Carolina	5
Louisiana	2		Tennessee	3
Maine	3		Texas	6
Maryland	1		Vermont	1
Massachusetts	4		Virginia	2
Michigan	2		Washington	3
Minnesota	2		Wisconsin	1
	9			
	1		TOTAL: 127	

Table 2
Fire Fighter Death Rate by Region
1987

Region	Number of Fatalities	Number of Fire Ground Deaths	Fire Ground Death Rate per 100,000 Fires
Northeast	44	16	3.10
Northcentral	25	10	1.91
South	36	13	1.43
West	22	14	3.66
Nationwide	127	53	2.27

rural or urban, and one must assign a fire fighter death as urban or rural based on the particular community in which he was operating when fatally injured.

Based on these rules, the following patterns were found and are shown with available patterns for the general population and for the population of fire fighters specifically in local fire departments:

	Urban	Rural	Total
Total 1987 fire fighter fatalities	79 (62%)	48 (38%)	127 (100%)
Suburban location	23	8	31
Local fire department only*	78 (70%)	34 (30%)	112 (100%)
U.S. population (1980)	74%	26%	100%
U.S. fire fighters (1986), total**	58%	42%	1 0 0 %
U.S. fire fighters (1986), career**	98%	2%	100%
U.S. fire fighters (1986), volun.**	47%	53%	100%

In 1986, the distribution of fire fighter fatalities from local fire departments was closer to the distribution of fire fighters from local fire departments than to the distribution of the whole U.S. population.*** The conclusions suggested by this finding were that the the risk of dying, as measured by local fire fighter deaths per 100,000 local fire fighters, is nearly the same in urban and rural areas. The result was surprising, however, since rural areas are generally served by volunteer fire fighters, who

*   Excludes one military service fire fighter killed in an urban location and 14 Federal, state and contract fire fighters killed in rural locations.

**   "U.S. Fire Department Profile Through 1986," Quincy, Massachusetts: National Fire Protection Association, Fire Analysis Division, November 1987. All percentages are for fire fighters in local fire departments.

** Note that the classification of fire fighters into urban and rural is based strictly on the population protected by the fire department and not on metropolitan area considerations. However, if fire fighter fatalities were similarly classified, the distribution would shift by at most two percentage points, so the points here are not affected.

typically would average fewer work (exposure) hours per year than career fire fighters and who would therefore be expected to have a lower risk of death per 100,000 fire fighters, as a separate career/volunteer analysis did show.

For 1987, however, the distribution of fire fighter fatalities from local fire departments (70 percent urban, 30 percent rural> is closer to the distribution of the whole U.S. population (74 percent urban, 26 percent rural) than to the distribution of fire fighters from local fire departments (58 percent urban, 42 percent rural). This suggests that, for 1987, urban fire fighters faced a greater risk of dying than fire fighters in rural areas. This is not unexpected since urban areas are generally served by career fire fighters who average more hours of exposure per year than volunteer fire fighters and therefore would be expected to have a higher risk of death per 100,000 fire fighters. This finding is supported by the fact that career fire fighters had 2.5 times the fire fighter death rate of volunteers in 1987. (Volunteers accounted for 64 of the 112 local fire fighter fatalities, or 57 percent, versus 77 percent of the local fire fighters, according to 1986 figures. Therefore the risk index for volunteers would be .57/.77 = .74 versus .43/.23 = 1.87 for career fire fighters.)

Since the results have changed from one year to the next, this will be worth rechecking in the future using data from multiple years.

III. TRAINING FATALITIES 1978-1987

In the ten-year period from 1978 through 1987, 58 fire fighters died while training. As shown in Figure 13, for the first nine years of the period, the number of deaths annually ranged between three and seven, but in 1987, the number increased sharply -- to 15 deaths.

The three activities that accounted for most of the deaths were physical fitness training or testing (12 deaths), apparatus and equipment drills (12 deaths) and live fire training (11 deaths). This last activity has received significant attention in the past year since it is the activity that resulted in the deaths of five participants in 1987. Three of the victims were killed in one fire when they were caught on the first floor of an old farmhouse by fire progress and became trapped after retreating to the second floor of the structure. They all died of soot and smoke inhalation6. In a separate training incident, a fire fighter died of asphyxiation after depleting his air supply while attempting to exit the burning building. The fifth victim in 1987 was struck by a car and fatally injured while directing traffic at a live fire training exercise.

The other fire fighters who died as a result of live fire training exercises include three who died of heart attacks, two who died of asphyxiation in one incident after being trapped in a flashover and one who died of burns when a home heating oil tank on the outside of the fire building exploded.

All twelve of the fire fighters who died during physical fitness training or testing were career fire fighters. Their activities at the time of injury were jogging, running, lifting weights and playing basketball. All suffered heart attacks.

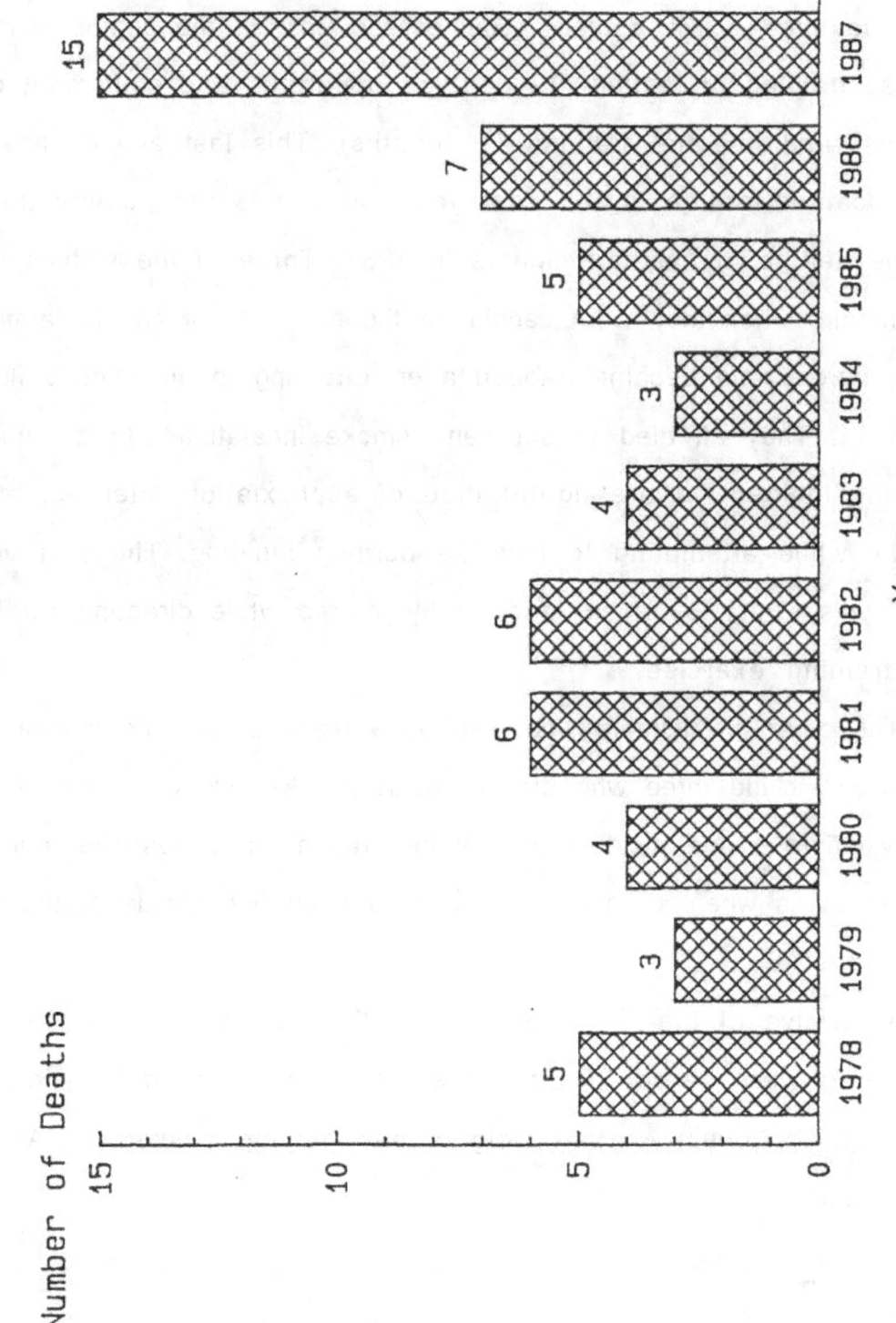

Figure 13
U.S. Fire Fighter Training Deaths
1978 – 1987

The other training activities where deaths occurred were classes and meetings (five deaths); smoke drills (four deaths); disaster and hazardous material drills (three deaths); underwater training (three deaths); preparation for training, including travel and setup (two deaths); high-rise drill (one death); and other unspecified training activities (five deaths).

Figure 14 shows the cause of injuries in training deaths. The most frequently reported cause of fatal injury was overexertion or stress (36 deaths). Eight deaths occurred when fire fighters were caught or trapped -- six by fire progress and one each by explosion and underwater objects. The seven deaths that resulted from being struck by or coming into contact with an object included five in motor vehicle collisions, one to contact with water and one electrocution. Five fire fighters died as a result of falls -- two from ladders, one from emergency apparatus, one tripped and fell on pavement, and one in a hole. Two fire fighters died as a result of exposure to smoke.

As shown in Figure 15, the categories for the nature of fatal injury were heart attack (32 deaths), internal trauma (8 deaths), asphyxiation (7 deaths), stroke (4 deaths), crushing (2 deaths) and one each to burns, drowning, respiratory arrest, electrical shock and aneurysm. Medical documentation was available for 12 of the fire fighters who died of heart attacks. Five were reported to have had severe arteriosclerotic heart disease, three had evidence of prior heart attacks or bypass surgery, two had hypertension, one had diabetes and one was reported to be fatigued at the time of the incident.

The fire fighters who died in training incidents ranged in age from 16 to 64 years. Thirty-four were career fire fighters and 24 were volunteer fire fighters.

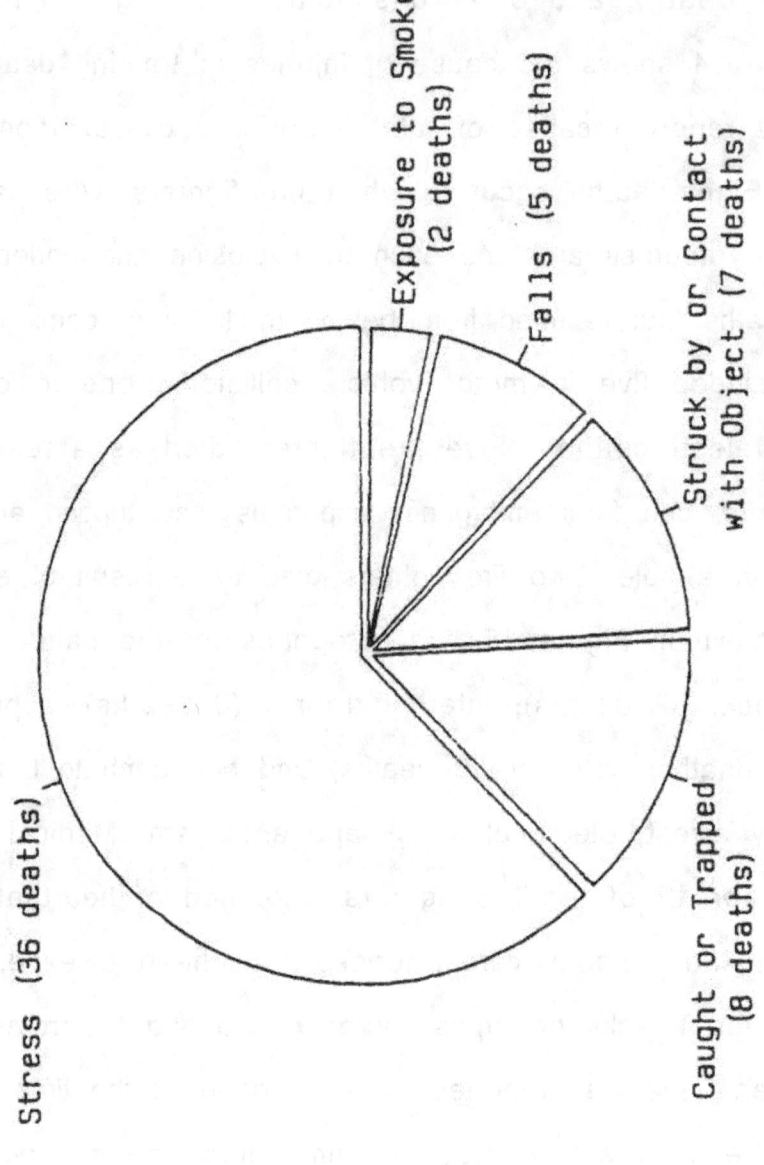

Figure 14
U.S. Fire Fighter Training Deaths
by Cause of Injury
1978 - 1987

Exposure to Smoke
(2 deaths)

Falls (5 deaths)

Struck by or Contact
with Object (7 deaths)

Caught or Trapped
(8 deaths)

Stress (36 deaths)

Figure 15

U.S. Fire Fighter Training Deaths

1978 - 1987

by Nature of Fatal Injury

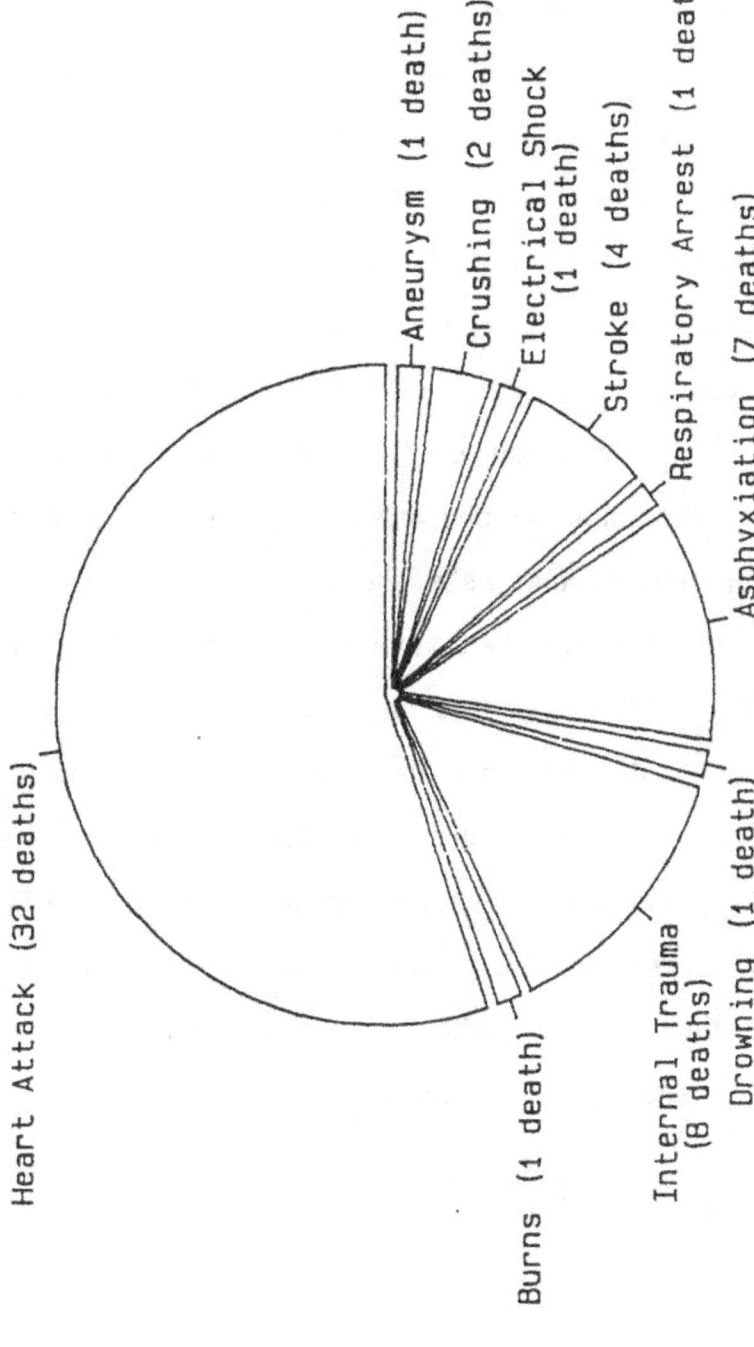

Aneurysm (1 death)

Crushing (2 deaths)

Electrical Shock (1 death)

Stroke (4 deaths)

Respiratory Arrest (1 death)

Asphyxiation (7 deaths)

Heart Attack (32 deaths)

Burns (1 death)

Internal Trauma (8 deaths)

Drowning (1 death)

Training accounted for less than five percent of all fire fighter deaths over the ten-year period but almost 12 percent of the deaths in 1987. Hopefully, we will find that this was a single-year phenomenon but it serves as an important reminder that although training is an essential part of fire fighting it can be dangerous.

Fire fighter fatalities involving training are indicative of the need for attention to safety, good command practices and appropriate standards on the training ground as well as the fire ground.

In this section, live fire training was highlighted as a specific problem area. Live fire training provides a high level of realism and simulation of actual fire ground conditions. However, such training exercises also pose many of the hazards encountered during actual emergencies. As a result, live fire training drills must be planned with great care and must be closely supervised to ensure that risks are kept to a minimum. A fatal live training incident in 1982 led NFPA's Committee of Fire Service Training to develop the standard, NFPA 1403, <u>Standard on Live Fire Training Evolutions in Structures.</u> One way to avoid future live fire training casualty incidents is through application of this standard along with good fire ground command practices.

All training exercises must be carefully planned and adequate precautions taken to minimize the risk to all participants. When the training involves particularly hazardous situations, extra supervision and precautions are necessary to ensure fire fighter safety.

IV. WILDLAND FIRE FATALITIES 1978-1987

During the period from 1978 through 1987, 1,278 fire fighters died in the line of duty. Of these 1,278 fire fighters, 147 (or 11.5 percent) died as a result of wildland fires. (For this analysis, the term "wildland" is used to include forest, brush and grass fires.) Approximately 17 percent of all fire ground deaths occurred in wildland fires. As shown in Figure 16, the number of deaths was generally between 12 and 17 per year with the exception of six deaths in 1982 and 22 in 1987. The sharp increase in deaths from 14 in 1986 to 22 in 1987 prompted this special analysis on fatalities related to wildland fires.

Three-quarters of the deaths (111) occurred during fire suppression activities. Their distribution by region is shown in Figure 17. The remaining 36 deaths occurred when fire fighters were responding to or returning from such fires. For the 66 wildland fires for which cause of the fire was reported, 28 were of incendiary or suspicious origin, 15 were due to misuse of heat (12 due to inadequate control of open fire, two due to children playing and one to a misfired tear gas cannister), eight were due to lightning and the rest to various other factors.

The 147 victims included 12 chief officers, 15 company officers and 120 fire fighters. They ranged in age from 17 to 87, with a median age of 42 years. Three of the victims were women.

The breakdown of causes of fatal injury is shown in Figure 18. The largest proportion of deaths during fire suppression activities were due to stress. These 40 deaths include 16 due to physical exertion on the fire ground. The next major category was contact with or exposure to object.

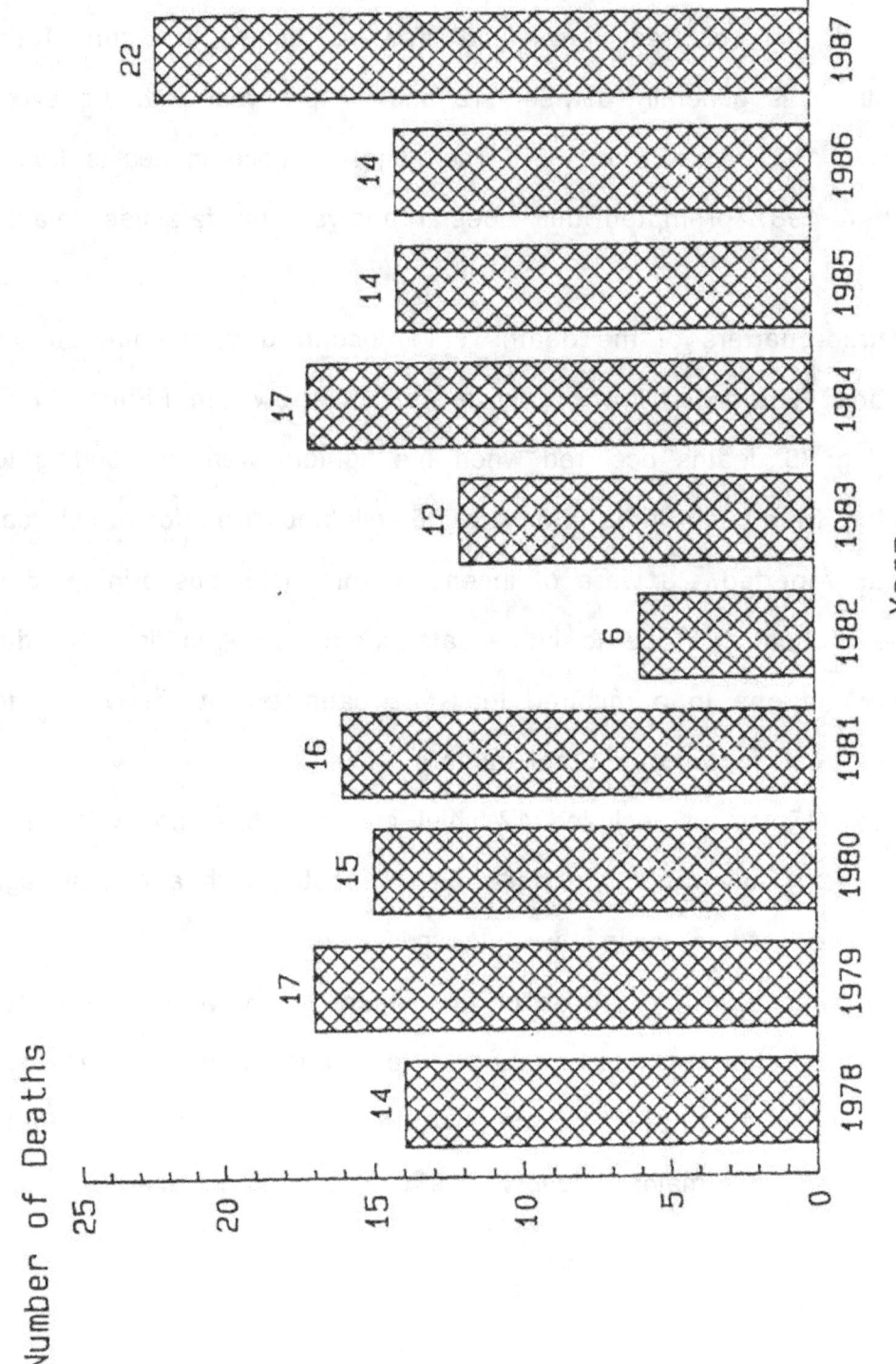

Figure 16
U.S. Fire Fighter Deaths
in Wildland Fires
1978 - 1987

Figure 17
Wildland Fire Ground Fatalities by Region
1978 – 1987

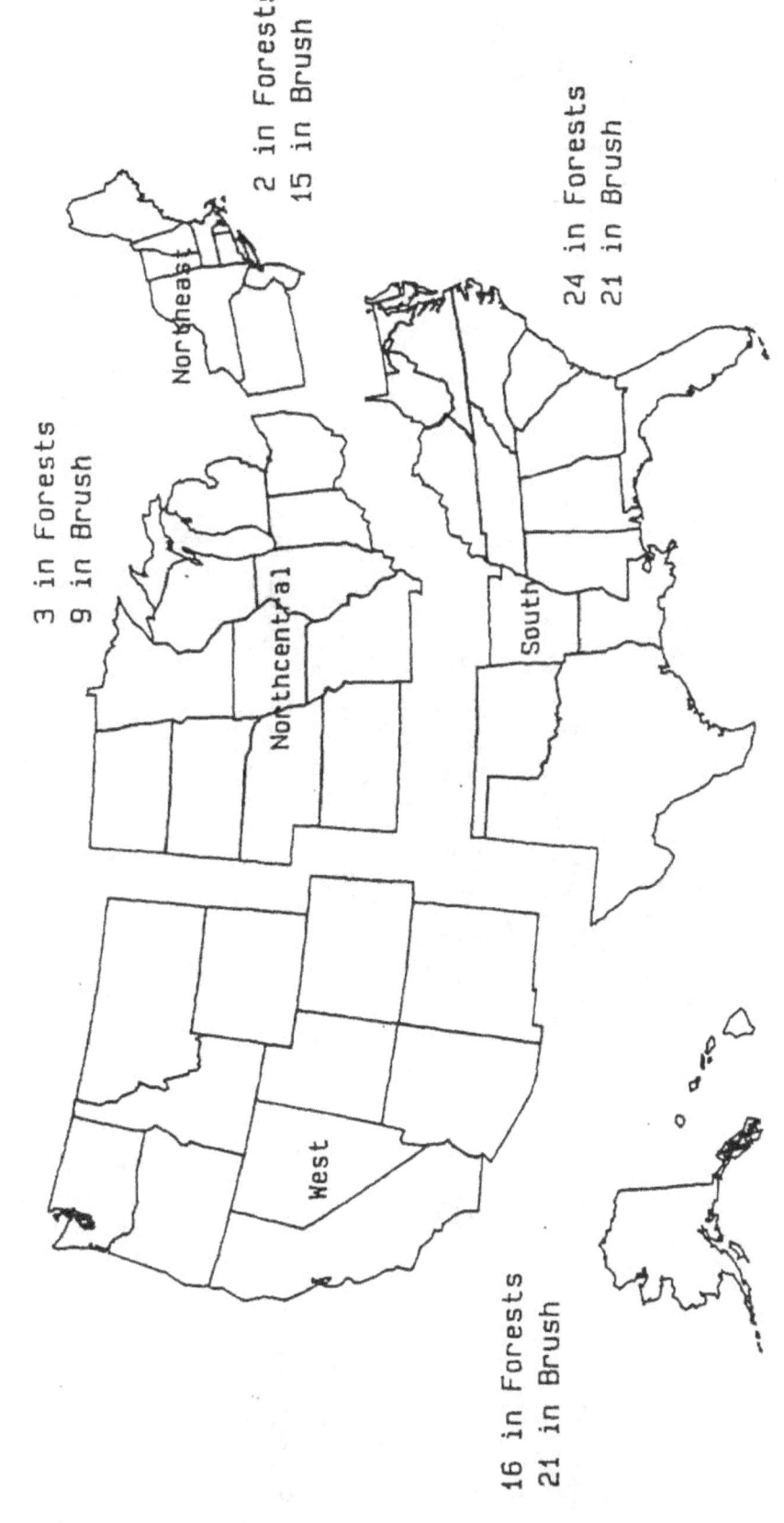

3 in Forests
9 in Brush

Northcentral

Northeast

2 in Forests
15 in Brush

South

24 in Forests
21 in Brush

West

16 in Forests
21 in Brush

Figure 18

Fire Fighter Fatalities in Wildland Fires

by Cause of Fatal Injury

1978 - 1987

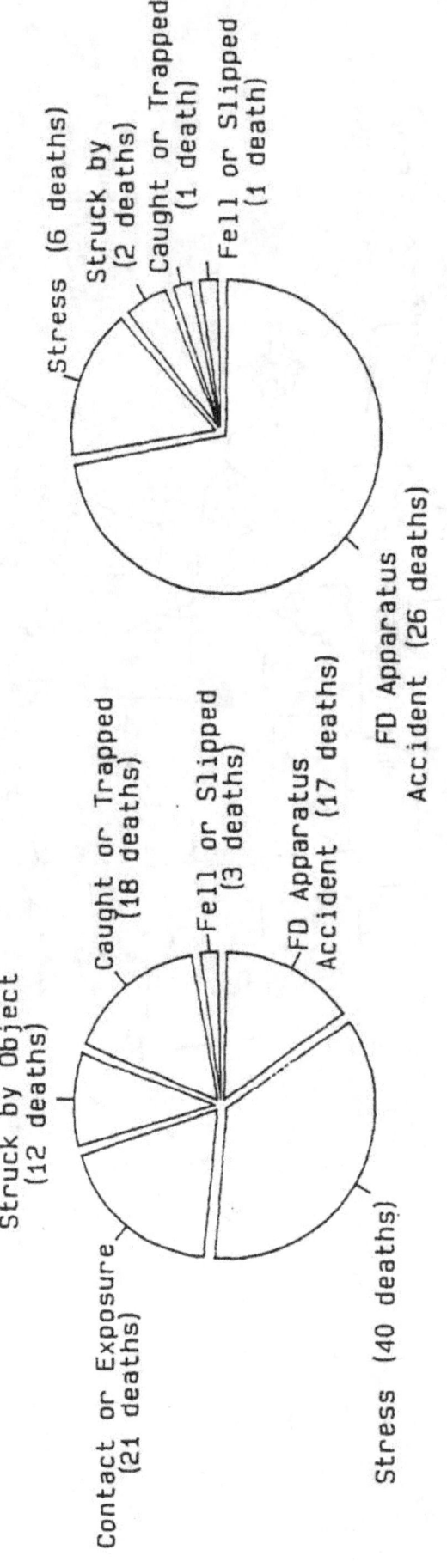

These 21 deaths include eight due to exposure to electricity, eight due to exposure to smoke, four due to exposure to fire and one due to a lightning strike. Three of the fire fighters who were electrocuted came in contact with downed powerlines; three were killed when they touched guy wires that had been energized from contact with power lines; and two made contact with their vehicles that had been energized. Two contributing factors for these electrocutions were tall vegetation and darkness.

The next major category was caught or trapped - 17 by fire progress and one by falling objects. Fourteen of those 17 fire fighters overrun by fires died of burns; the other three were asphyxiated. Another seventeen fire fighters were killed in fire department apparatus accidents during fire suppression activities (including 12 in aircraft crashes) and 12 others were struck by vehicles or falling objects. The remaining three fire ground fatalities fell - two from emergency apparatus and one in a hole.

Of the fire fighters who died while responding to or returning from wildland fires, 26 were killed in accidents involving fire department apparatus (including 5 in aircraft crashes).

The nature of fatal injuries is shown in Table 3. For this analysis, the fire ground deaths were broken down by type of department - municipal departments (career or volunteer) or forestry agencies. A profile of the 111 fire ground victims shows that 64 were members of municipal fire departments (52 belonged to volunteer departments and 12 to career departments). The other 47 fire fighters were career, seasonal or contract employees of state and federal forestry agencies. As the table shows, heart attacks accounted for more than half of the deaths of municipal fire fighters during fire suppression activities, while most of the deaths to state and federal employees were due to internal trauma and burns.

State and federal forestry officials believe that their rigid fitness requirements account for the low proportion of heart attack deaths, which seems to be supported by these findings. Of the 21 municipal heart attack victims for whom medical documentation was available, 12 had had prior heart attacks or bypass surgery, 6 had severe arteriosclerotic heart disease, one was fatigued, one had diabetes, and one had hypertension. The municipal volunteer fire fighters who suffered heart attacks ranged in age from 44 to 71 years with a median age of 55. The career municipal fire fighters ranged in age from 48 to 61 with a median age of 53. In contrast, of the five forestry fire fighters who died of heart attacks, one had severe arteriosclerotic heart disease, one had hypertension and no medical documentation was available on the other three. They ranged in age from 33 to 57 with a median age of 38 years.

As far as the other types of injuries suffered on the fire ground are concerned, increased use of fire shelters may result in a reduction in fatal burns and smoke inhalation deaths and the use of safer aircraft may reduce the number of deaths due to aircraft crashes during suppression activities.

As mentioned earlier in this report, in 1987, for the first time in the eleven years of NFPA's comprehensive study of fire fighter deaths, the largest proportion of fire ground deaths occurred in wildland fires. In addition to those 17 deaths, another five deaths occurred while responding to or returning from such fires. Seven of the 17 deaths, and two of the responding/returning deaths, occurred in California during a one-month period when the states of California and Oregon were plagued by a series of wildland fires. One incident that claimed the lives of three contractors to the U.S. Department of the Interior occurred when an air tanker crashed while dropping fire retardant.

In another incident, two U.S. Army contractors were killed when their aircraft crashed as they were dropping fire retardant on a range fire near the White Sands New Mexico missile range. Six other fire fighters suffered fatal

Table 3
U.S.. Wildland Fire Fighter Fatalities
by Nature of Fatal Injury
1978-1987

	Fire Ground Deaths			
	Federal and State Wildland Agencies	Municipal Volunteer	Career	Deaths While Responding/Returning
Asphyxiation	4	0	2	1
Thermal Burns	12	5	2	0
Cardiac Arrest	5	31	7	6
Crushing	3	2	1	6
Drowning	0	1	0	0
Hemorrhaging, Bleeding	0	2	0	0
Internal Trauma	19	5	0	23
Electrical Shock	3	6	0	0
Heat Stroke	1	0	0	0
	47	5 2	12	36

heart attacks while operating at wildland fires. Two fire fighters who were cutting trees at two separate forest fires were killed when they were struck by falling tree sections. A fire fighter died of burns and smoke inhalation when he was caught by rapid fire progress while cutting vegetation at a forest fire. One fire fighter was struck and killed by a motorcycle while operating at a forest fire. Another was killed when the bulldozer he was operating rolled over. In another incident, a fire fighter fell from the extended bumper area of a brush vehicle and was run over by the front wheels.

V. SMOKE INHALATION AND SMOKE EXPOSURE DEATHS, 1978-1987

Exposure to smoke and its aftereffects is an obvious hazard of fire fighting. This brief analysis is confined to fire ground deaths due either to asphyxiation due to smoke inhalation or heart attacks related to smoke exposure -- the types of injuries relevant to the use of self-contained breathing apparatus (SCBA).

There were 161 smoke inhalation and smoke exposure heart attack deaths during fire ground activities from 1978 through 1987 (24.6 percent of the total number of fire ground deaths). Approximately three quarters of them (122 deaths) were due to asphyxiation; the remaining 39 deaths were heart attacks. The distribution of these deaths by year is shown in Figure 19. In order to draw any useful conclusions regarding the incidence of smoke-related deaths, it is helpful to look at the two categories of fatal injury separately.

For the 39 heart attack deaths, smoke exposure was explicitly mentioned in death certificates or autopsy reports as a major contributing factor in the heart attack. In many cases, carboxyhemoglobin (COHB) levels were also provided. It is interesting, however, to look at some factors also mentioned in the medical documentation for these fatalities. Fourteen of the 39 victims had had prior heart attacks or bypass surgery. Another 12 had severe arteriosclerotic heart disease (at least 50% occlusion). No information on prior physical condition was provided for the other 13 victims. The victims ranged in age from 29 to 71 with a median age of 46 years. Twenty-four of the victims were career fire fighters; the other 15 were volunteers.

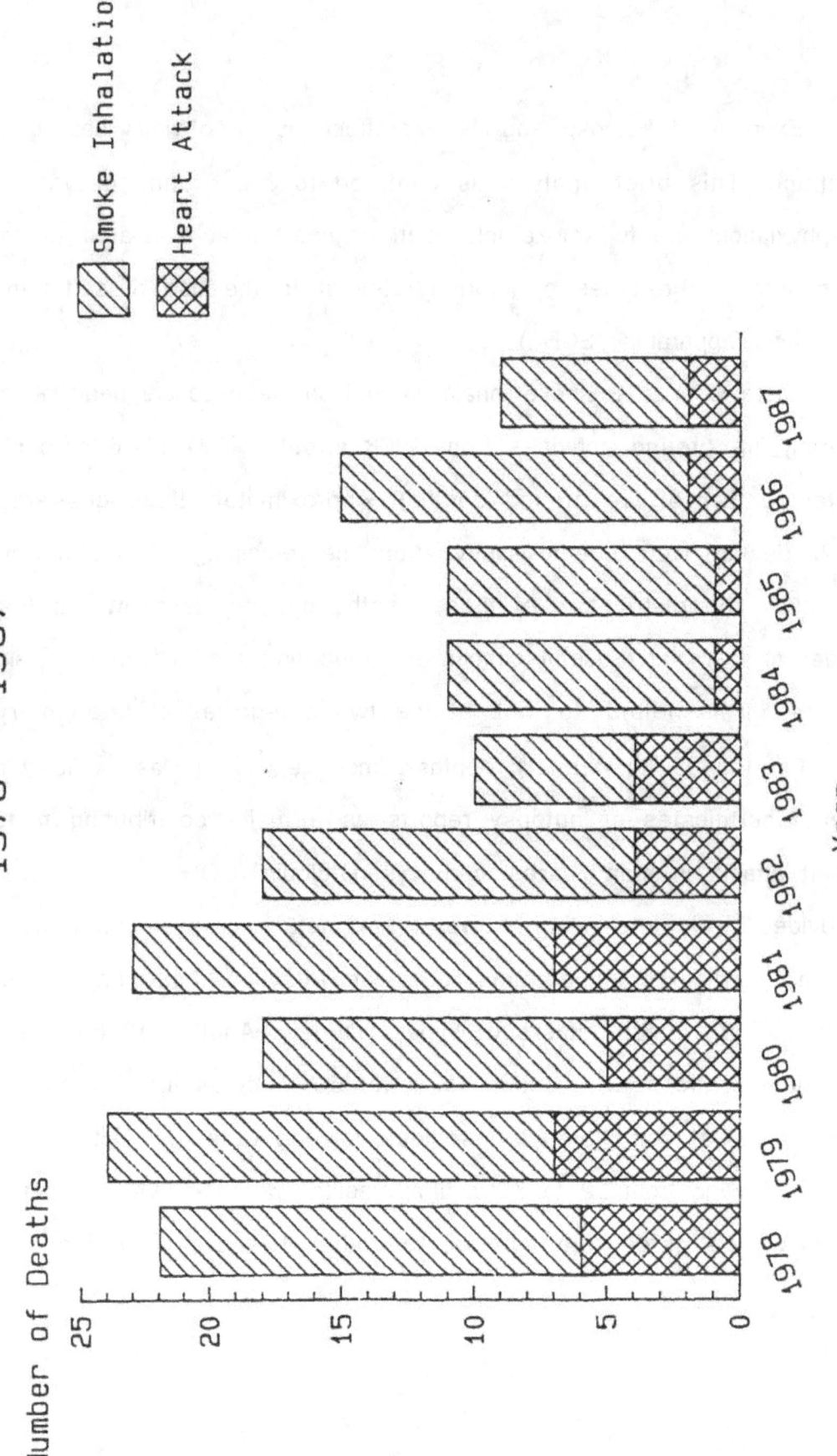

Figure 19
Smoke Inhalation and Smoke Exposure
Fire Ground Deaths
1978 - 1987

Of these 39 heart attack victims, 30 were not wearing SCBA at the time they were stricken. Most of them reportedly had not been using SCBA at any time during the fire. Some had been wearing SCBA during fire fighting operations, but removed their masks during overhaul. In some instances, the wearing of SCBA was reported, but the high COHb levels reported made the actual use of the equipment doubtful.

The 122 fire fighters who died of smoke inhalation had a somewhat different profile. They ranged in age from 17 to 67 with a median age of 32 years. Of the 52 for whom medical documentation was available, 45 were rested, five were reportedly fatigued, one was impaired by alcohol, and one suffered from severe arteriosclerotic heart disease. Seventy-eight of the victims were career fire fighters; the other 44 were volunteers.

The causes of the smoke inhalation deaths are listed in Table 4 and shown in Figure 20. The most frequently reported cause was lost inside structures (28 deaths). Other major categories were exposure to smoke (25 deaths), caught by flashover (16 deaths), trapped by collapsing roof (16 deaths), caught by fire progress (15 deaths) and trapped by floor collapse (15 deaths).

Of the 28 fire fighters who were lost inside structures, 22 had SCBA but ran out of air, three had removed their SCBA or facepiece, one had equipment that leaked and no information on SCBA was available for the other two. Of the 25 who died as a result of exposure to smoke, 12 were not using SCBA (five were fighting fires in their own homes, four were members of a fire brigade and three were fighting grass or brush fires), seven ran out of air, two had removed their SCBA, one had a broken facepiece and no information on SCBA was available for the other three.

There were 16 other fire fighters without SCBA who died of smoke inhalation. Ten were exposed as a result of roof collapse (eight were on the roof, two were inside), four were trapped by fire progress during wildland fires and two were trapped by flashover during fires in their own homes. Eleven others who were using SCBA had the equipment burned, melted, severed, knocked off or otherwise detached as a result of the circumstances that led to their fatal injuries.

It appears from the decrease in the number of smoke inhalation and smoke exposure deaths over the ten-year period, that the improvements in the design and testing of SCBA which were strongly advocated in the late seventies and early eighties seem to have had a positive impact on the problem. The other (and maybe most significant) factor is the increased awareness among fire fighters of the need to use SCBA and the adoption by many departments of mandatory guidelines for the use of SCBA. In order to further reduce the occurrence of such deaths, efforts must continue to ensure the safety of the fire fighters operating inside structures including maintaining the integrity of their escape route, using and heeding the warnings of low air alarms, the utilization of PASS devices, and, of course, proper fire ground management.

Table 4
Causes of Smoke Inhalation Deaths
1978 - 1987

Causes	Number of Deaths
Caught or Trapped	
Lost inside	28
Flashover	16
Collapsing roof	16
Fire progress	15
Collapsing floor	12
Backdraft	2
Collapsing ceiling	1
Exposure to	
Smoke	25
Struck by	
Bales of hay	3
Collapsing. roof	1
Falls	
Through hole in roof	2
Through hole in floor	1

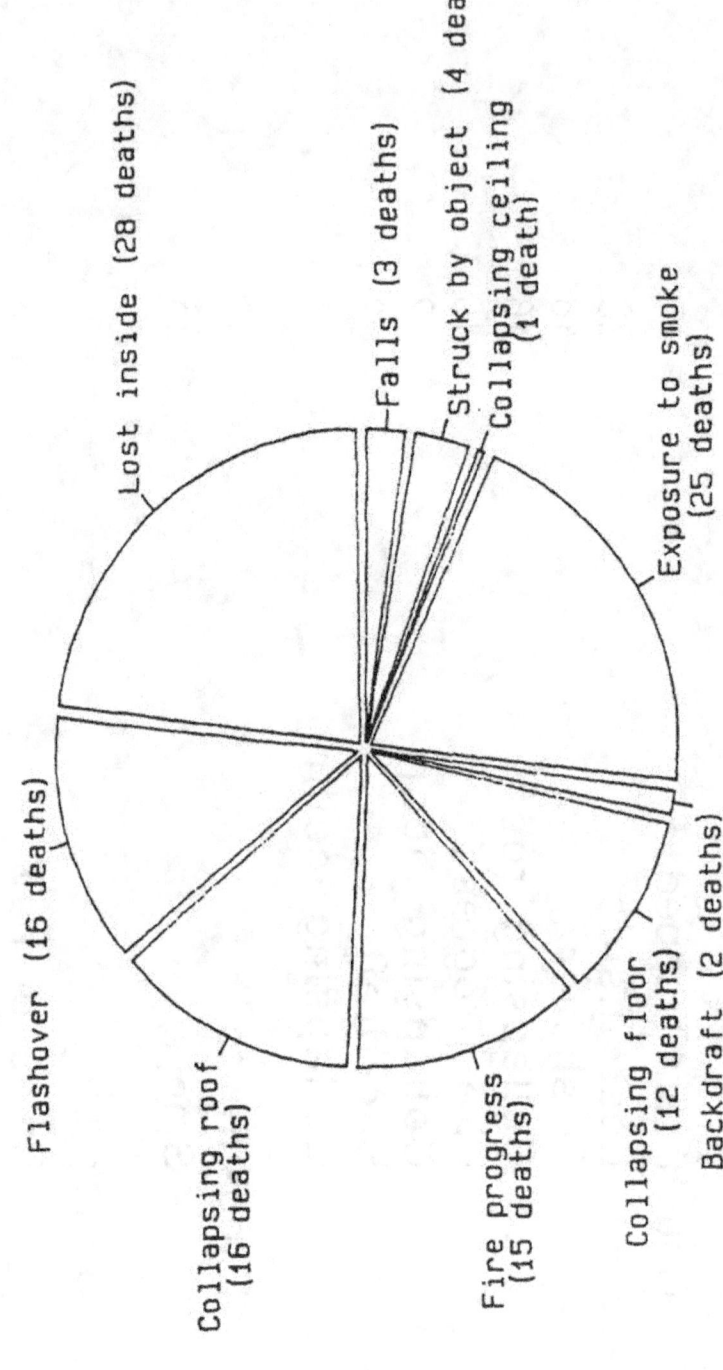

Figure 20
Causes of Smoke Inhalation Deaths
1978 – 1987

Lost inside (28 deaths)

Falls (3 deaths)

Struck by object (4 deaths)

Collapsing ceiling
(1 death)

Exposure to smoke
(25 deaths)

Flashover (16 deaths)

Collapsing roof
(16 deaths)

Fire progress
(15 deaths)

Collapsing floor
(12 deaths)

Backdraft (2 deaths)

VI CONCLUSIONS AND RECOMMENDATIONS

Although there has been some reduction in the number of fire fighter fatalities since the late 1970's, a plateau seems to have been reached and more aggressive efforts must be made to reduce fatalities further. The purpose of this report is to document some of the findings from the 1987 experience and the last decade's experience in order to use the lessons learned to improve fire fighter safety.

This report analyzed three areas of particular concern: deaths during training activities, deaths related to wildland fires, and smoke inhalation and smoke exposure deaths. The analyses of training and wildland deaths are prompted by the experience in 1987 and serve to highlight two areas of the fire fighter fatality problem that, while not representing the largest aspects of the problem, are significant.

Training is an essential fire fighter activity, but care must be taken to ensure the safety of all participants. Recognized safe practices for conducting live fire training drills in structures are outlined in NFPA 1403. It is important to note, however, that most training-related fire fighter deaths (40 of 58 in 1978-87, or 69 percent) did not involve such inherently hazardous activities as live fire training, smoke drills, or underwater training. In fact, most of the deaths during training reflect the factors seen in the larger group of fire fighter deaths, i.e., heart attacks, often associated with serious pre-existing medical problems, and the hazards of operating apparatus, specifically vehicle accidents and falls. An examination of training-related deaths serves, as much as anything else, to reinforce the need for physical fitness screening and training and safer apparatus-handling practices.

Although the 1987 wildland fire experience may be a single-year phenomenon, it did focus attention on an area where attempts to decrease losses, including more rigid fitness requirements and testing, have shown signs of success and other strategies, such as the use of fire shelters, have the potential for further reduction in losses. Vehicle-handling problems seem to need attention here, too. There also are special problems in wildland fires that can lead to fire fighters becoming lost or isolated. These special risks warrant continued attention to command, control, and communications issues.

The smoke inhalation and smoke exposure analysis was an effort to followup on studies on SCBA-related deaths done in the early eighties. The analysis showed that most of the victims who had been using SCBA ran out of air, usually after becoming lost inside or trapped. Most of the victims without SCBA were operating at wildland fires, on roofs or at fires in their own homes. This analysis helps to illustrate the importance of proper use of SCBA and proper management of the fire ground to ensure that fire fighters are properly accounted for and are not operating in areas where they become lost or subject to building collapse. Fire fighters must maintain contact with each other and ensure that they have access to an exit while operating in a structure.

Areas that merit additional study include investigation into the medical histories of heart attack victims for whom documentation of prior heart problems was not available; an examination of the cancer deaths reported to PSOB, whether they qualified for benefits or not;' and a comprehensive analysis of factors in motor vehicle accidents while responding to or returning from alarms. Two relevant topics in the latter category are fatal accidents involving volunteer fire fighters responding in their own vehicles and fatal and non-fatal accidents involving emergency vehicles.

To an individual fire department, the death of a fire fighter can appear to be a random and extremely rare event. However, a look at the national experience can provide valuable lessons to all departments. Changes in operating procedures and attitudes must be made to improve fire fighter safety.

REFERENCES

1. "National Traumatic Occupational Fatalities, 1980-1984," National Institute for Occupational Safety and Health, Division of Safety Research, Morgantown, WV, June 11, 1987.

2. Michael J. Karter, Jr., "Taking the Measure of the fire Service," Fire Command, Vol. 52, No. 7, (July 1985).

3. NFPA Fire Analysis and Research Division, "Estimated Fire Fighters in the U.S., 1985," unpublished, February 1987.

4. Michael J. Karter, Jr., "U.S. Fire Loss in 1987," Fire Journal, Vol. 82, No. 5, (September 1988).

5. The four regions as defined by the U.S. Census Bureau included the following 50 states and the District of Columbia:
    Northeast: Connecticut, Maine, Massachusetts, New Hampshire, New Jersey, New York, Pennsylvania, Rhose Island, and Vermont.
    Northcentral: Illinois, Indiana, Iowa, Kansas, Michigan, Minnesota, Missouri, Nebraska, North Dakota, Ohio, South Dakota, and Wisconsin.
    South: Alabama, Arkansas, Delaware, District of Columbia, Florida, Georgia, Kentucky, Louisiana, Maryland, Mississippi, North Carolina, Oklahoma, South Carolina, Tennessee, Texas, Virginia, and West Virginia.
    West: Alaska, Arizona, California, Colorado, Hawaii, Idaho, Montana, Nevada, New Mexico, Oregon, Utah, Washingon, and Wyoming.

6. Thomas J. Klem, "Fatal Live Fire Training Incident," Fire Command, Vol. 55, No. 5, (May 1988).